Radioactive Radiations and their Biological Effects

Peter Lindenfeld
Rutgers University

An Issue-Oriented Module

Published and distributed by
American Association of Physics Teachers

Radioactive Radiations and Their Biological Effects

Revised 1982
Third Printing, March 1986
Cover Design by Karen S. Garvin

For order information, write to:
American Association of Physics Teachers
Publications Department
5110 Roanoke Place, Suite 101
College Park, MD 20740, U.S.A.

ISBN 0-917853-08-3

Contents

Preface

This booklet is intended to supplement an introductory physics course with material which is seldom discussed in elementary textbooks, but which is of great current interest and importance.

Part 2 deals with radiation measurement, units, and natural background radiation. Part 3 discusses in some detail the biological damage from ionizing radiation, and seeks to relate the effects to the elementary interactions with which the student is familiar. The section headed "Readings, Quotations, and Bibliography" is meant to be the basis for class discussion. The goal here, as in the whole module, is twofold. First, some basic facts are introduced. But beyond this the student should become aware of the importance of the subject and of the nature of the controversies around it. It is my hope that this background will enable the student to contribute more knowledgeably and more effectively as an informed citizen.

It is assumed that the student will already be familiar with the basic facts about atoms and nuclei as they are usually presented in elementary physics courses. Some of the relevant topics appear in the Appendix, "A Summary of Some Properties of Nuclei, Nuclear Radiation, and Reactors." Also included are two problem sets, as well as discussion questions.

The subject of this booklet is somewhat different from most of those which traditionally find their way into elementary physics. It bridges physical science, biological science, and social science. Discussion questions, essays, book reviews, and literature summaries may be particularly appropriate here.

The temptation to include more material was very strong. But while excellent and much more complete and detailed books exist, there seemed to be little that was likely to find its way into an introductory course.

I believe that this material is too important to be left to more specialized courses. I was motivated by the conviction that we have a responsibility to teach physics as a living science, essential for an understanding of some of the most important current issues.

Acknowledgments

I would like to thank my colleagues on the Issue-Oriented Modules Advisory Board of the AAPT for providing the impetus to write this material. I am indebted to them and to the reviewers whom they chose, for their helpful comments. In my attempts to learn some biology I have profited from conversations with my friend and neighbor, Mrs. Greta Sander.

I would like to thank the Environmental Protection Agency for Figure 1, Doubleday and Company for Figure 2, Argonne National Laboratory for Figures 3, 4, 5, and 6, the U.S. Energy Research and Development Administration for Figure 7, N. W. Ayer ABH International for Figure 8, and the Public Service Electric and Gas Company for Figure 9.

Peter Lindenfeld
Rutgers University
August, 1983

1. *Introduction*

Did you know that each of our bodies contains some radioactivity? Every minute about a quarter of a million of the atoms in your body transform into atoms of other elements as their nuclei radiate some of their energy.

Is that a large number? Well, that depends. Large compared to what? It looks like a big number, but as a fraction of the total number of atoms in the body it is extremely small. How about the radiation which is emitted—can it have some effect on the rest of the body? Yes it can, and the nature of this effect is what we wish to discuss and investigate.

Part of our concern is that we are exposed to radioactive radiation from inside and outside our bodies all the time. It is obviously interesting to consider what the effect of this radiation may be. But there is another more urgent reason for this study: the amount of radiation to which we are exposed is increased by man-made sources. Nuclear reactors are more and more widely used to generate electricity. Nuclear weapons are made and tested. X rays and radioisotopes are used for medical diagnosis and therapy. Each of these sources affects individuals as well as the whole population. Much, but not all, is known about the effects which such radiation has on us. There are important benefits. There are also dangers.

Twenty years ago the debate about radiation was primarily over the testing of nuclear weapons. Today it is mostly about using nuclear reactors for power generation. All of us are affected, in ways which are not completely known and understood, and over which we usually have little control. It is not surprising that the discussion is often emotional.

Our plan in this booklet is to study the sources and the effects of nuclear radiation in some detail. We will then be in a position to consider some of the questions and controversies which surround this subject. The object is not to provide answers. We will see, rather, that there are situations in which thoughtful and informed people can come to different conclusions.

2. Natural Background Radiation and Its Measurement

Cosmic Rays

Most of the radiation to which we are exposed comes from outside our bodies. First there are cosmic rays. Their name implies radiation which comes from outside the earth. Actually the radiation reaching us consists almost entirely of particles which are produced in the upper atmosphere in nuclear reactions caused by the true (or "primary") cosmic rays. Near the surface of the earth about three quarters of these particles are muons (or mu mesons) resulting from the decay of pions (or pi mesons) which in turn are among the products of the nuclear reactions. The rest are mostly electrons and photons. At sea level the rays arrive at the rate of about one per cm^2 per minute, but the rate is approximately doubled for each 2000 m increase in altitude.

Natural Radioactivity; the Becquerel and the Curie

Another source of nuclear radiation is the radioactive material in our environment. There are small amounts of uranium and thorium in the ground as well as in the building materials which we use. The relatively common element potassium contains a small fraction which is radioactive. The half-lives of these substances are about a billion years or more. We believe them to be left over from the time, several billion years ago, when the elements were formed, with all possible stable and unstable isotopes. The ones which are still around are either stable or have sufficiently long half-lives to have survived.

Some radioactive materials are being created now. Uranium and thorium transform radioactively into other elements, which in turn decay until, finally, the end product is stable lead. One of the steps in the succession of decay products is the gas radon. A special case is radioactive carbon (^{14}C) which is created in the atmosphere by nuclear reactions of cosmic rays on nitrogen.

The amount of radioactive material may be specified by giving the activity, that is, the rate at which the nuclei decay. One disintegration per second is called a becquerel (Bq). A curie (Ci) is 3.7×10^{10} Bq. If there are N radioactive nuclei, the activity is equal to λN, where λ is called the decay constant. The decay constant is related to the half-life, $T_{1/2}$, by $\lambda T_{1/2} = 0.693$.

Measurement of Radiation Effects

How much radiation are we exposed to? Before we answer this question let us consider just what we might try to measure. We could ask about the number of particles that hit us. We have given an estimate for the cosmic rays as one per cm^2 per minute at sea level. In doing so we have ignored the vastly larger number of neutrinos which we believe to rain down on us. That is because the neutrinos pass right through us with a negligible chance of leaving even the slightest trace. It is clear that we are not so much interested in the number of rays as in the effect which they might have on us.

That leaves us with a more difficult question: How does the radiation affect us? The biological results are complex, but the basic physical mechanism which is responsible for them is straightforward: Charged particles passing through matter exert electrical forces (described by Coulomb's law) on the charged particles of which matter is composed. A sufficiently fast particle can transfer some of its energy to an atom. The atom is excited to a higher energy state, often so high that it loses one of its electrons. An atom that loses an electron is said to be ionized. The radiation is called ionizing radiation.

Even radiation which is not composed of charged particles may cause ionization. It cannot do so directly, but it will ionize indirectly if it gives some or all of its energy to a charged particle.

Alpha rays (helium nuclei) and beta rays (electrons) are charged particles and ionize directly. Gamma rays and x rays are electromagnetic rays (photons) and do not. They may, however, give their energy to electrons through the photoelectric effect, the Compton effect, or pair production.

Ionization, Exposure; the Roentgen

One of the qualities that we might wish to measure is therefore the amount of ionization which a given quantity of radiation can produce. This is called the "exposure." It may be specified as the amount of charge which is liberated in the form of electrons in a given amount of air at standard temperature and pressure. In SI units the exposure is given in coulombs of electrons per kilogram of air. Another unit is the roentgen (R). It produces 2.58×10^{-4} coulombs of electrons per kilogram of air (at standard temperature and pressure, "STP," i.e. 0°C and atmospheric pressure).

Absorbed Energy; Absorbed Dose; the Gray and the Rad

The amount of ionization which is actually produced will depend on the nature of the material in which the radiation is absorbed. In air at STP one ion pair is produced for each 34 eV of absorbed energy. In other materials the energy per ion pair is different.

The most important and at the same time the simplest quantity which characterizes radiation effects is the amount of energy which is absorbed per unit mass of the absorbing material. This is called the "absorbed dose." The SI unit of absorbed dose is the gray (Gy), equal to one joule per kilogram. The rad is equal to 10^{-2} Gy.

Note that the exposure and the absorbed dose are quite different quantities: the exposure describes a property of the radiation; the absorbed dose describes something that happens when the radiation is absorbed by a particular material.

Biological Response to Radiation; Dose Equivalent; the Sievert and the Rem; Relative Biological Effects

The question remains whether the absorbed energy is really a good measure of the biological effects which the radiation produces. It turns out that it's not bad, but we cannot be surprised that something so complex and varied as the effect of radiation on living matter is only roughly related to any single physical quantity.

We may also have to consider the rate at which the energy is absorbed (the "dose rate" in Gy/s or rads/yr) and the volume in which the absorption takes place. The influence of these variables may differ depending on whether we are talking about local tissue damage, the production of a particular disease, or genetic effects. Different individuals may be affected differently depending on their age and physical condition.

To measure the biological effect of a given amount of radiation we need therefore yet another unit: we wish to define it in such a way that the same amount of biological damage occurs regardless of the nature of the radiation causing it. The quantity which is measured is called the "dose equivalent," but when no ambiguity is likely to arise it is usually just called the dose.

Let's start with the easy part: it turns out, on the basis of experimental observations, that for the most common kinds of radiation the biological effect depends, to a good approximation, only on the absorbed energy. There is then no difference between the dose equivalent and the absorbed dose. This is the case for x rays, for gamma rays, and for most beta rays. (The beta rays have to have an energy spectrum which extends to more than 30 keV.)

The starting point for the definition of the units for dose equivalent is therefore to make them equal to the units of absorbed dose so long as only x rays, gamma rays, and sufficiently energetic beta rays are involved. The SI unit for dose equivalent is the sievert (Sv). The rem is equal to 10^{-2} Sv. So long as we are dealing only with x rays, gamma rays, and energetic beta rays, these units are equal to the gray and the rad.

If the equivalence between absorbed dose and dose equivalent held for all ionizing radiation, there would, of course, be no need to define a new quantity and new units. Experiments show, however, that for some kinds of radiation a given amount of absorbed energy can lead to much greater biological damage than for x rays, gamma rays, and beta rays. For these radiations the Sv will not be equal to the Gy, and the rem will not be equal to the rad.

Only biological experiments can determine the relation between dose equivalent (H) and absorbed dose (D). Their ratio, in any particular experiment, is called the "relative biological effectiveness," or RBE, so that $H = D \times RBE$.

We have now left the realm of physics. The dose equivalent depends on the kind, the amount, and the rate of the radiation; on the nature of the organism which is being irradiated, on the organism's age, its sex, state of health, and surroundings, and on the particular biological effect which is being considered. The bridge from physics to biology is a difficult one to cross!

Ionization and absorbed energy are physical quantities which can be defined and measured with comparative ease. But now we have to deal with the interactions of large and complicated molecules in a varied and incompletely known environment. We have to measure large-scale effects whose relation to the molecular processes is often very imperfectly known.

Linear Energy Transfer; the Quality Factor

Fortunately we can get back to physics with a few approximations which are usually quite acceptable. We assume that it does not matter which particular biological effect we are considering, or which particular organism is being irradiated. We also ignore, at least to start with, differences in dose rate.

We are then left only with the properties of the radiation itself. Experiments show that the RBE depends only to a minor extent on the factors which we are neglecting, but that it depends strongly on the kind of radiation, and sometimes also on the energy of the radiation. For alpha particles, for example, the RBE is approximately equal to 10. In other words, a given amount of absorbed energy resulting from alpha radiation causes about ten times as much biological damage as the same absorbed from gamma radiation (which is assigned an RBE of 1).

The main reason for the difference in RBE between the different radiations is well known and understood. It is that some particles lose their energy after traversing a relatively short distance, so that the ionized atoms along their trails are close together. The damage which is produced is then greater than for particles of the same initial energy which cause ionizations which are further apart.

We define the "linear energy transfer" (LET) as the amount of energy per unit distance which is lost by the radiation along its trail. It may be measured in J/m, but is more usually described in kiloelectron volts per micron (keV/μm).

The LET goes up with the mass of the particle which produces it. Alpha particles lose their energy in much less distance than the much lighter beta particles and their RBE is correspondingly greater.

By ignoring all influences on the RBE except for the linear energy transfer, we have come back to the "simple" realm of physics. The RBE was supposed to describe the true biological effect. We can now define a factor which does so only approximately, but which depends only on the nature of the radiation.

We simply assign a value of this new factor to each kind of radiation, depending only on the LET of the radiation. The value which is assigned is equal to the best known value of the RBE for the radiation. The new factor is called the quality factor (Q). Table I shows the values which have been adopted.

Table I. Quality Factor of Different Radiations

	Q
x rays, gamma rays	1
beta rays whose maximum energy is greater than 0.03 MeV	1
beta rays whose maximum energy is less than 0.03 MeV (e.g. those from tritium)	1.7
neutrons and protons up to 10 MeV	10
naturally occurring alpha particles	10
heavy recoil nuclei	20

As long as the internationally assigned and adopted value of the quality factor for alpha particles remains 10, that is the value which is to be used in calculations which relate dose equivalent and absorbed dose for alpha particles. It is quite possible, however, that a particular biological experiment may lead to an RBE for alpha particles which differs from 10. If a number of new experiments were to lead to an RBE significantly different from 10, the assigned value of the quality factor would presumably be changed by the appropriate international committee.

The Q is, as well as we can make it so, equal to the RBE. The two factors are logically quite different, but it will not surprise you that the difference is often ignored. What happens is that the term RBE is used even when Q is meant. This does not normally bother anyone, even though a weighty committee (the "RBE Committee") issued a weighty report, in which it warned that "it is extremely necessary that a clear distinction should be made . . ." [*Health Physics,* **Vol. 9,** p. 357 (1963)].

Background Dose Estimates

At last we can summarize our knowledge of the natural background radiation absorbed per person by giving the number of millirems per year from the various sources (1 mrem $= 10^{-3}$ rem). (See Table II.)

Table II. Natural Background Radiation

cosmic rays	28 mrems/yr
external radioactivity	26 mrems/yr
internal radioactivity	28 mrems/yr
total	82 mrems/yr

These estimates are for the average dose to the whole body in the United States. They are from a report which we will refer to for most of the quantitative estimates of radiation effects. It is called *The Effects on Populations of Exposure to Low Levels of Ionizing Radiation,* and was made by the Advisory Committee on the Biological Effects of Ionizing Radiations of the National Academy of Sciences in 1980. It is usually referred to as BEIR III. (An earlier version, BEIR I, was issued in 1972.)

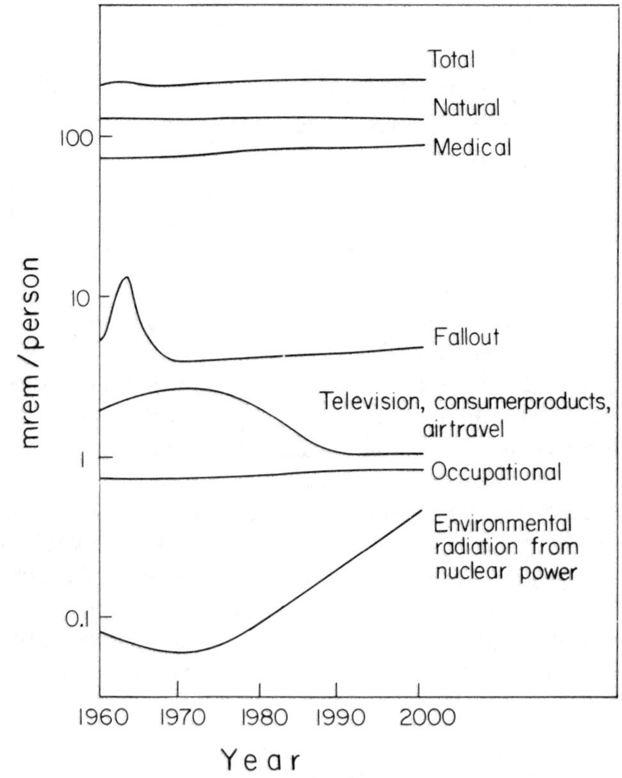

Figure 1. Average estimated yearly dose per person in the United States.

The numbers are obviously subject to considerable uncertainty. You may see other estimates, made by other people, which differ, sometimes by as much as 30%. Remember also that they are averages, and that there are variations from place to place. Finally, different parts of the body may be exposed to different amounts of external radiation because they will be shielded differently. They may also be exposed differently to internal radiation because the distribution of radioactive substances in the body is not uniform.

Figure 1 shows an extrapolation into the future for the radiation doses to be expected in the next decades. It is taken from a report called *Estimates of Ionizing Radiation Doses in the United States 1960-2000,* made for the U.S. Environmental Protection Agency in 1972. It gives the average expected yearly dose in mrem per person in the United States.

Problem Set 1

(Note: Other examples and problems are given in the Appendix, "A Summary of Some Properties of Nuclei, Nuclear Radiation, and Reactors.")

The examples have intermediate steps worked out to three significant figures. The final answer can be rounded off to give a smaller number of significant figures if this is appropriate.

Conversion factors and other information can be found in Table A-I.

Example:

Strontium 90 has a half-life of 28 years. It emits beta rays whose maximum energy is 0.55 MeV. The product nucleus is Yttrium 90 which decays in a relatively short time ($T_{1/2}=64h$) by beta-emission with a maximum energy of 2.27 MeV.

The net result (after a few days elapse, creating an equilibrium amount of ^{90}Y) is the same as if ^{90}Sr emitted both beta rays with its half-life of 28 years. About 60% of the decay energy is carried off by the neutrinos which accompany the beta rays.

(a) What is the activity of one gram of ^{90}Sr in equilibrium with its daughter ^{90}Y?
(b) What is the energy given to the beta rays by one gram of ^{90}Sr in one half-life?
(c) What is the number of watts of heat power produced in the surroundings by one gram of ^{90}Sr?

Answers:

(a) The gram-atomic mass is (approximately) 90. That is, the number of atoms in 90 grams of ^{90}Sr is Avogadro's number,

$N_o = 6.02 \times 10^{23}$. The number of atoms in one gram is $N = N_o/90 = 6.69 \times 10^{21}$.

The decay constant is

$$\lambda = 0.693/T_{1/2} = (0.693/28)\ yr^{-1}$$
$$= (0.693/28)/3.16 \times 10^7\ s^{-1}$$
$$= 7.83 \times 10^{-10}\ s^{-1}$$

The activity is λN

$$= (6.69 \times 10^{21})(7.83 \times 10^{-10})\ Bq$$
$$= 5.24 \times 10^{12}\ Bq$$
$$= 5.24 \times 10^{12}/3.7 \times 10^{10}\ Ci$$
$$= 142\ Ci$$

(b) In each disintegration the energy given to the beta rays is 40% of $(2.27+0.55)$ MeV or $0.4 \times 2.82 \times 1.6 \times 10^{-13} J = 1.80 \times 10^{-13} J$. In one half-life the number of atoms which disintegrate is $(N_o/90)/2$ so that the energy given to the beta rays is:

$$(1.80 \times 10^{-13})(6.023 \times 10^{23}/180) = 6.04 \times 10^8 J.$$

(c) $(5.24 \times 10^{12}$ disintegrations/s$)(1.80 \times 10^{-13})$ J/disintegration $=$ 0.943 W.

Problems

1. Ionizing radiation is absorbed in 1 kg of air at STP, and produces an absorbed dose of one rad.
   (a) How many ion pairs are created?
   (b) What is the total charge of the electrons in this number of ion pairs?
   (c) What is the exposure in roentgens corresponding to this charge?
   (d) What would be the percent error if it is assumed that 1 R produces 1 rad in the air?

2. Radium (^{226}Ra) has a half-life of 1620 years. An alpha particle with a kinetic energy of about 5 MeV is emitted when a radium nucleus decays to become a nucleus of radon.
   (a) Write down the nuclear reaction equation which describes this decay.
   (b) What is the number of atoms in one microgram (10^{-6}g) of radium?
   (c) What is the total amount of energy which is emitted during one half-life by a source of radium whose initial mass is one microgram?

3. The potassium content of the human body is about 0.21%, and 0.012% of the potassium is the radioactive isotope ^{40}K. The half-life of ^{40}K is 1.25×10^9 years. Each decay releases 1.3 MeV, and about 40% of this amount is absorbed by the body. (All the neutrinos and some of the gamma rays leave without being absorbed.)

(a) What is the number of ^{40}K atoms in a 70 kg person?

(b) What is the decay constant λ of ^{40}K in s^{-1}?

(c) What is the number of ^{40}K atoms which disintegrate per second in the body of the 70 kg person? What is the activity in microcuries (10^{-6} Ci)?

(d) What is the energy which is absorbed in the body per second? What is the absorbed dose in one year?

4. A 70 kg person ingests one microcurie of ^{14}C. Calculate the absorbed energy and the absorbed dose in one week. (Assume that none of the ^{14}C is eliminated, and that it is uniformly distributed throughout the body. ^{14}C emits beta rays with a maximum energy of 0.16 MeV. About 60% of the released energy is in the form of neutrinos which leave without being absorbed.)

Figure 2. Mme. Curie and her daughter Irene, 1925. From *Madame Curie* by Eve Curie, translated by Vincent Sheean. © 1937 by Doubleday & Company, Inc. Reproduced by permission of Doubleday & Company, Inc.

3. Biological Effects of Radiation

The Discovery of Radioactivity and Its Biological Effects

Roentgen discovered x rays in 1895. The discovery of the radioactive radiations by Becquerel was inspired by Roentgen's work and followed soon afterwards. Pierre and Marie Curie discovered polonium and radium in 1898. The realization that both kinds of radiations could cause "burns" came almost immediately. Becquerel and Marie Curie developed accidental skin lesions and Pierre Curie exposed himself purposely to study the effect.

The existence of delayed effects was not realized, and insufficient precautions were taken. Both Marie Curie and her daughter Irene (who with her husband discovered artificial radioactivity) died of leukemia (a type of cancer which affects the blood). In the biography of her mother her other daughter, Eve, quotes what she calls the verdict of science: "Marie Curie can be counted among the eventual victims of the radioactive bodies which she and her husband discovered."

Supposed benefits were ascribed to radium, and preparations containing it were widely sold. Today much is known about the effects of ionizing radiation, yet it is surprising to see how much uncertainty remains and how many questions remain unanswered.

The Three Stages of Radiation Effects

Let us consider briefly how ionizing radiation can affect biological processes. There is no great mystery about the general nature of the effect. Atoms are held to one another to form molecules by electrical forces. On the atomic scale chemical reactions are just rearrangements of electrons. The molecules of living beings are large and complicated; their electrical forces are delicately balanced. Clearly the removal of an electron will affect the balance drastically, and is likely to lead to a change in the molecule's structure. The molecules, or the fragments which remain, may no longer be able to fulfill their former functions. Some biological damage will have occurred.

The sequence which we have described consists of three stages. The first is the physical stage in which energy is absorbed, most likely by a single electron.

The second stage involves the molecule and its immediate surroundings. The structure of one or more molecules changes, in other words, chemical reactions take place.

Only a fraction of a second has passed when the third stage begins. Now the biological processes may be affected by the altered composition. Repair mechanisms may come into play to undo the damage. If they are unsuccessful and if the damage is sufficiently serious, the molecule or the cell will no longer be able to play its normal part in the biological processes. Eventually the whole organism may be affected, but perhaps not until months or years have passed.

Genetic Effects; Chromosome Damage and Mutation

No molecule, and therefore no living cell, can completely resist the disruptive effect of ionizing radiation. The most sensitive process is reproduction, and the molecule which is the most important target is DNA.

The DNA molecule carries the information which is necessary for the replication of body cells (called "somatic" cells) by division. In the "genetic" cells the DNA carries and transmits the characteristics which are inherited through sexual reproduction.

Let us first look at the essentials of the mechanism of sexual reproduction which are relevant to our purpose. Each cell contains chromosomes. Each chromosome consists of genes. The genes are made of DNA and are the carriers of inherited characteristics. In human beings the somatic cells have 23 pairs of chromosomes and the genetic cells have 23 single chromosomes. On fertilization the two sex cells each contribute their single chromosomes to a new set of pairs. The number of genes is unknown, but has been estimated to be about 40,000 in each human cell.

Two kinds of damage are known to occur in the genetic material. One kind involves the chromosome as a whole. There may be the wrong number, for example. Or—and this is more likely to be produced by ionizing radiation—a chromosome may break. The broken pieces often reattach, but in ways which may or may not leave the chromosome's function impaired.

The second kind of damage is to a particular gene and is called a gene mutation, a point mutation, or simply a mutation. Several results may occur: one is that the change is so drastic that inheritance of the gene makes development impossible. Another possibility is that the mutation causes an observable change in the individual that inherits the gene, regardless of the corresponding gene in the second chromosome of the pair. In that case the mutation is said to be dominant. Hundreds of diseases and abnormalities are known to result from this cause.

The third possibility is that the mutation will only cause an observable result if corresponding genes in both chromosomes of the pair carry exactly the same mutation. Such a mutation is said to be recessive. It can be inherited for many generations without making itself apparent. It forms part of the "gene pool" of the population. Only when a particular fertilization finds it next to another mutuation of the same kind can it affect the individual that has inherited it.

The presence of potentially damaging recessive mutations in the gene pool is referred to as the "genetic load" on the population. It has been estimated that most human sex cells are likely to contain recessive lethal genes. As a result, between 5 and 50% of fertilized egg cells are estimated to be "lost" genetically either before or soon after conception, often so soon that the mother never becomes aware that she was briefly pregnant.

It seems that the largest number of mutations are those which have minor effects. They may be almost, but not quite, recessive, and their cumulative effect is therefore particularly hard to assess.

Mutations are essential for the process of evolution. A mutation may cause an improvement and make it possible for the organism that inherits it to cope better with its environment. If that happens, the improved organisms are likely to succeed better, and gradually displace the less well adapted ones.

On the other hand mutations are random, unplanned changes. Our present level of evolution makes us enormously complicated. The chances that a random change will be beneficial are extremely small. An increase in the rate at which mutations occur can only be damaging.

Acute Somatic Effects

Any of the functions of the somatic cells may be impaired by radiation. Just as in the genetic cells, however, the most important effects seem to be those which result from damage to DNA so that they affect the reproductive mechanism of the cell.

Some kinds of cells divide much more frequently than others, and it is in these cells that radiation-induced damage is most likely to become apparent. Among these are the blood-forming cells of the bone marrow and the cells which line the intestinal wall. The radiation sensitivity of these two kinds of cells is mainly responsible for the acute (short-term) effects of large doses of radiation.

The extra sensitivity of dividing cells also accounts for the fact that the embryo is particularly susceptible to radiation damage.

The radiation doses which cause short-term effects are fairly well known. A dose to the whole body of 25-100 rems causes blood changes, but the affected individual will not normally be aware of them. "Radiation sickness" with malaise and vomiting will result from 100-300 rems, but the person is likely to recover. Higher doses lead to widespread damage throughout the body, particularly to the gastrointestinal tract and to the bone marrow. With a dose of 500 rems the chance of recovery is about 50%. Above 600 rems death is almost certain.

Cancer

Cancer occurs when something goes wrong with the way in which cells reproduce, so that uncontrolled growth occurs. The relation between cancer and radiation has two aspects. On the one hand cancer cells, because they are constantly dividing and reproducing, are strongly affected by radiation. This is the basis for radiation therapy. On the other hand the initiation of cancer ("carcinogenesis") can result from radiation. Again we are confronted with the effect of radiation on the cells' reproductive properties.

It has proved extremely difficult to study radiation-induced cancer in humans. The main reason for this is that most of the time cancer is initiated in some other way. Certain chemicals, for example, are carcinogenic. A second reason why cancer studies are so difficult is that it often takes years after the initiating event before cancer becomes apparent. The incidence of leukemia among atomic bomb survivors reached a peak six years after the explosion. In some cases it took twenty years for the disease to develop.

Estimates of Damage from Low Dose-Rates; Extrapolation from High Dose-Rates

As far as we know, only a small fraction of cancerous diseases and genetic defects is caused by radiation under normal circumstances. The size of this fraction is not known, although many attempts have been made to estimate it.

How can one make such an estimate? In principle there is a straightforward method: a group of individuals is irradiated with a certain amount of additional radiation over and above the normal background, and the resulting defects are measured.

The difficulties with this method are obvious. We do not make such experiments on human beings. Nevertheless, some data on cancer exist for atomic bomb survivors, for patients who have been given radiation treatment, and for people who have been irradiated as a result of their work before the hazards were understood.

An estimate of the cancers produced by a given dose can be made, but only in the rare cases where the exposure has been much larger than that of background radiation. How can we use this estimate to tell us how many cancers are produced by radiation if it is absorbed at the vastly smaller rate of background radiation? How can we extrapolate the graph of dose against effect from large to small doses?

This is a crucial question. We wish to know what the effects of radiation are at or near background level. But the experimental information comes only from very much greater exposures.

There are two kinds of evidence which can help us to know how to estimate the effects of low doses. One is our knowledge about the molecular and cellular processes which lead to somatic and genetic damage, and those which lead to repair and recovery. The second kind of evidence comes from direct observation of the relation between radiation doses and the effects which they produce. For human beings there is only very limited evidence on radiation-induced cancer and virtually none on genetic effects. A great deal of work has been done on animals, but the results at low dose rates are difficult to establish. In addition it is not clear how to scale the results so as to make them applicable to humans. In any case no experimental results exist for dose rates as small as background. We must extrapolate from results at higher rates.

In the absence of convincing evidence the method of extrapolation has become highly controversial. BEIR I (in 1972) concluded that dose and effect should be proportional for both cancer and genetic effects. A crucial result of this assumption is that a given dose produces the same effect regardless of the time over which the dose is accumulated. In other words, the effect is independent of the dose rate.

A second report was withdrawn almost as soon as it was issued in 1977. Finally in 1980, a third report (BEIR III) appeared. The data which the committee had to go on were not signficantly different from those of BEIR I. This time, however, the recommendations of the committee were different and much more complicated. For genetic effects the linear extrapolation was retained. For cancer, on the other hand, there was no agreement. The majority decided on a relation between dose and effect which had both a linear term and a quadratic term. But there were minority reports, one of which argued in favor of the linear relationship, while another came out for a purely quadratic form.

In the extrapolation from high to low doses the linear relation leads to a higher estimate of the risk from low-dose radiation. The committee's majority considers it to define an upper limit for almost all cases, while the purely quadratic relation is likely to give a lower limit.

It is important to note that none of these extrapolation methods imply a "threshold" dose below which no biological effect occurs.

Quantitative Estimates of Low-dose Effects

In view of our very limited knowledge of the effects of low doses, and the controversy about the method of estimating them, it is perhaps useful to emphasize that for most cases the various estimates are, in fact, reasonably close to each other. We will list some of the estimates of both reports, BEIR I and BEIR III.

For genetic effects the evidence is poor and the uncertainty large. Both reports use linear extrapolation methods and come to about the same conclusions. BEIR III puts the results as follows: One rem of total dose to all parents leads to 5-75 additional serious genetic disorders per million live births in the next generation.

A one-rem dose to each generation will eventually lead to 60-1100 additional serious genetic disorders in each million births. (The normal incidence of such disorders is about 107,000 per million births.)

The "doubling dose," which multiplies the number of natural mutations by a factor of two, is estimated to be between 50 and 250 rems.

For cancer the evidence is greater, but the two reports differ in their method of analysis. The BEIR I results are given for a dose of 5 rems accumulated over a time of 30 years. This corresponds roughly to a doubling of the present rate of natural and man-made radiation.

This amount is estimated to lead to between 3,000 and 15,000 additional cancer deaths per year in the United States. The most likely number is 6,000 deaths per year, or 1.8×10^5 deaths in the 30-year period. For a population of 200 million, each receiving a dose of 5 rems, there will be 10^9 "person-rems," or 1.8×10^{-4} deaths per person-rem.

BEIR III estimates the excess cancer mortality from a single dose of 10 rads in a population of one million to be 770, and for a continuous dose of 1 rad/yr throughout life, as 4,700. These numbers are roughly half as large as those derived by using linear extrapolation. Purely quadratic extrapolation leads to numbers smaller still by a factor of about ten. (In the United States cancer accounts for about 16% of all deaths.)

These estimates are for low-LET radiation, i.e. for radiation whose RBE and Q are equal to one. For high-LET radiation (neutrons, alpha particles, etc.) BEIR III suggests that linear extrapolation be used. The estimates are then the same as in BEIR I.

Choices

Although quantitative estimates of the biological effects of ionizing radiation are not very precise, there is no doubt that exposure to such radiation can produce a variety of deleterious results. If we expose ourselves to more than the inescapable minimum it should, presumably, be only for very good reasons.

Sometimes we have a personal choice: We can choose not to work in the nuclear industry. But can we refuse a doctor's recommendation to have an x ray? What if our government decides to explode an atomic bomb over a remote island in the South Pacific? What if a foreign government does it?

It is not our objective here to argue for one or another choice in any given situation. We merely try to consider some of the factors that must be taken into account. The case studies which follow illustrate some of the controversies which have arisen, and some of the different points of view.

Problem Set 2

The number of cancer deaths from radiation is estimated in BEIR I to be 1.8×10^{-4} deaths per person-rem. The conclusions of BEIR III cannot be expressed so simply, but lead to results about half as large in some examples. In order to make simple estimates in these problems please use 10^{-4} cancer deaths per person-rem.

Example:

A radioactive cloud passes over New York and exposes 10 million people to radiation with an average dose of 20 mrems. What is the number of excess cancer deaths to be expected? By what fraction will these deaths increase the normal cancer mortality?

A dose of 0.02 rems for 10^7 persons corresponds to $(0.02)(10^7) = 2 \times 10^5$ person-rems. The number of excess cancer deaths is $(2 \times 10^5)(10^{-4}) = 20$.

Since cancer accounts for 16% of deaths in the U.S. 1.6×10^6 deaths from cancer are expected in the population of 10^7 persons. The fractional increase is therefore $20/1.6 \times 10^6$ or 1.25×10^{-5}, or 1.25×10^{-3} percent.

By assuming a constant effect per person-rem we have implicitly made a linear extrapolation. If quadratic extrapolation is appropriate our estimate is too high. The example demonstrates why the effects of low doses of radiation are so difficult to observe. The expected additional 20 cases of cancer are indistinguishable from the 1.6 million cases which occur in the absence of the extra radiation.

Problems:

1. Assume that everyone in the U.S. (2×10^8 persons) is exposed to an additional 5 mrem/year. What is the number of additional cancer deaths to be expected, per year?

2. The cosmic-ray intensity in Denver (pop. 5×10^5) is about three times that of the average for the U.S.
   (a) What is the number of rems of additional radiation in Denver per year, and per generation of 30 years?
   (b) What is the number of extra cancer deaths per year to be expected from the additional radiation?
   (c) What is the number of additional serious genetic disorders to be expected in Denver per million births?
   (d) Assume a doubling rate of 100 rems. What is the percent increase in the normal mutation rate attributable to the extra cosmic ray intensity in Denver?

3. Assume a doubling rate for mutations of 100 rems. What is the fraction of normal mutations which is attributable to background radiation during one generation of 30 years?

4. The average whole-body radiation from medical x rays is given in Table III as 72 mrems/yr. How many cancer deaths per year can be expected to result from this amount of radiation in the U.S.?

5. Approximately 2.16×10^6 people lived within 50 miles of the Three Mile Island nuclear-reactor accident on March 28, 1979. They received an average dose of 1 mrem from the radiation which resulted

from the accident during the following two weeks.

(a) What was the collective dose in person-rems?

(b) What is the expected number of cancer deaths resulting from this radiation?

6. What is the radiation dose for one person so that there will be a 10% probability that the person will eventually die from cancer as a result? Is it likely that cancer will develop as a result of this single dose?

4. *Readings, Quotations, and Bibliography*

Background Radiation

One of the difficulties in dealing with ionizing radiation is that we have no sense organs which can detect it directly. We have no built-in warning mechanisms similar to those which protect us from too much light, sound, or heat.

In an effort to make radiation seem more natural and acceptable, attention is often drawn to the fact that we are constantly exposed to background radiation. The preoccupation with this message often causes any mention of the possibilities of damage to be neglected.

Figure 3 shows a picture and its legend taken from a booklet called *The Natural Radiation Environment* by J. Kastner (U.S. Atomic Energy Commission, 1968).

The last section in the same booklet is headed "Relax and Enjoy It," and contains the following: "Will genetic damage by radiation alter homo sapiens? As we don't know, and there really isn't much we can do about it anyway, let's hope the changes, if any, will be for the better. Goodness knows we can certainly bear to be improved."

A very different perspective is given in another booklet of the series, called *The Genetic Effects of Radiation* by I. Asimov and T. Dobzhansky. They make clear that an increased mutation rate cannot be beneficial, and say: "What humanity must do, if additional radiation damage is absolutely necessary, is to take on as little of that added damage as possible, and not pretend that any direct benefits will be involved."

There's More to the Sun Than Meets the Eye

The girl in the picture is getting more than she bargained for in the form of external radiation exposure. Besides the ultraviolet photons from which she hopes to develop a tan (an effect inside the skin), she is being bombarded by even more penetrating radiation, both from the earth and from outer space. She can't feel it because there are no nerve endings deep within her body. Furthermore, there are few lasting effects since the body is pretty good at repairing this minor internal damage, which might be compared with a tiny sunburn.

Figure 3. A soothing view of radiation.

Here is a petulant remark made during a debate on nuclear power:

> It should be noted that temperature, like ionizing radiation, can cause genetic mutations and that as much as 50 percent of the mutations that occur normally in contemporary man might be due to the increase in testicular temperature caused by the male practice of wearing trousers. Although this observation appeared in the literature in 1957, I am unaware of any subsequent popular movement to prescribe kilts in place of the more mutagenic habit of dress of the American male. [Merrill Eisenbud, in *Nuclear Power and the Public* (University of Minnesota Press, 1970), p. 81.]

Medical Use of X Rays

The BEIR reports are among the best references for the material which we have been discussing. Both reports contain a wealth of information. Table III (from BEIR I) shows the estimates for the average whole-body dose per person in the United States in 1970.

Table III. Natural and Man-Made Radiation

Source	**Dose rate in mrems/yr**	
environmental		
natural	102	(BEIR III gives 82 mrems/yr. See Table II.)
fallout	4	
nuclear power	0.003	
medical		
diagnostic x rays	72	(abdominal dose)
occupational	0.8	
miscellaneous	2	
total	182	

There is no doubt that medical x rays are the largest source of man-made radiation. This realization has only very slowly led to more careful and considered use by the medical profession. Shielding and collimation can reduce unnecessary exposure significantly, particularly the genetically significant dose to the sex organs.

BEIR I says that "medical radiation exposure can and should be reduced considerably by limiting its use to clinically indicated procedures utilizing efficient exposure techniques and optimal operation of radiation equipment." This would seem to be an obvious and elementary guideline.

How widely it is disregarded has been described in an article by Claire Nader (Ralph Nader's sister) "The Dispute Over Safe Uses of X Rays in Medical Practice" *Health Physics,* **Vol. 29,** p. 181 (1975).

Spokesmen for medical professional organizations have consistently pooh-poohed warnings of possible risks, and strongly resisted any limitations on physicians' decisions to order x rays. Claire Nader quotes a typical reply to one of her brother's articles: "All qualified experts agree with me that the American people are not being endangered by present diagnostic medical and dental use of radiation. . . . Except in the early years after x rays were discovered in the last century, there have been no identifiable injuries, much less deaths, caused by the diagnostic use of x rays."

Typical exposures are 45 mR for a chest x ray and 1100 mR for a single dental film. Routine chest and dental x rays are widely made in the absence of specific indications of their necessity. Whether the risk that accompanies the radiation is justified by the possibility of early detection of disease and decay is not clear.

A particularly difficult choice concerns screening for breast cancer. An exposure of 1R is considered reasonable for mammography (x rays of the breast). On the other hand a recent article states that at the centers of the National Cancer Institute the dose per examination ranged between 0.3 and 6.5 rads. ["Breast Cancer: Second Thoughts About Routine Mammography," *Science,* **Vol. 193,** p. 555 (1976)] As a result of the balance between benefit and risk it is now suggested that such x rays be used without a specific indication only for women over 50 years old.

Gofman and Tamplin

An interesting aspect of the history of the BEIR reports is the role played by John Gofman and Arthur Tamplin. These two scientists, with substantial academic credentials, were on the staff of the Livermore Laboratory of the Atomic Energy Commission in 1969 when they called for a ten-fold reduction in the "allowable radiation dose" of 170 mrem/yr.

Gofman and Tamplin emphasized the likely results to be expected if everyone in the U.S. were actually exposed to this limiting dose. Their writing, published in two books and dozens of articles, is unfortunately in a style that is so polemical that many people are repelled by it. (The books are *Poisoned Power* and *Population Control Through Nuclear Pollution.*) Nevertheless there can be little doubt that Gofman and Tamplin were instrumental in causing widespread awareness of the radiation problem and a reassessment of radiation standards. In 1971 the AEC lowered its standards for emission of radioactive materials from nuclear reactors.

BEIR I devotes six pages to an analysis of Gofman and Tamplin's point of view. C. L. Comar, the chairman of the committee that wrote the report, says:

> Gofman and Tamplin deserve full credit for emphasizing the view that the possible somatic effects of low-level radiation exposure should not be ignored . . . [and] . . . full discredit . . . first for conveying . . . the idea that development of nuclear power . . . could lead to total average population exposure anywhere near 170 millirems per year . . . [and] second, for promoting the simplistic theme that if nuclear power is not developed, then society will be spared the associated or equivalent biological cost. (*Bulletin of the Atomic Scientists,* September, 1973)

The Radium Watch-Dial Painters

The following excerpt is from an article in the *Journal of the American Medical Association* for December 5, 1925, entitled "Some Unrecognized Dangers in the Use and Handling of Radioactive Substances," by Martland, Conlon, and Knef.

> This report is published now as a warning that when long-lived radioactive substances are introduced into the body, either by way of the gastrointestinal tract (as they were in these cases), or by way of intravenous injections for therapeutic effects (as is being advocated for the treatment of such conditions as gout, arthritis, arteriosclerosis, leukemia, and Hodgkin's disease), death may follow a long time after, from the effects of constant irradiation on the blood-forming centers.

This excerpt is from the first of a number of articles which describe the effects on a group of women who were employed to paint watch dials with luminous paint which contained radium. It was their normal practice to point the brushes with their lips. By 1950 Dr. Martland had recorded 41 deaths among the women, and described the bone destruction, bone cancer, and blood disorders which resulted from the presence of the radium. (See Figs. 4, 5, and 6.)

Although the use of radium as a supposedly healthful tonic decreased sharply after the first publication by Martland and his associates, a 1955 article states that ". . . the administration and taking of soluble radium salts has continued sporadically to the present time in the United States" (*Radiation,* by J. Schubert and R. E. Lapp, p. 114).

Fallout

Fallout from atomic bomb tests is seen to be the second largest source of man-made radiation. (See Table III and Fig. 1) The average dose had reached 13 mrem/yr in 1963 when the Test-Ban Treaty stopped atmospheric tests by most (but not all) countries. The present level is the result of long-lived fission products from the tests before 1963 as well as from the relatively small number of tests since then. This level is not expected to decrease significantly in the next decades.

Figure 4. Radium-dial painters in a watch factory in 1922

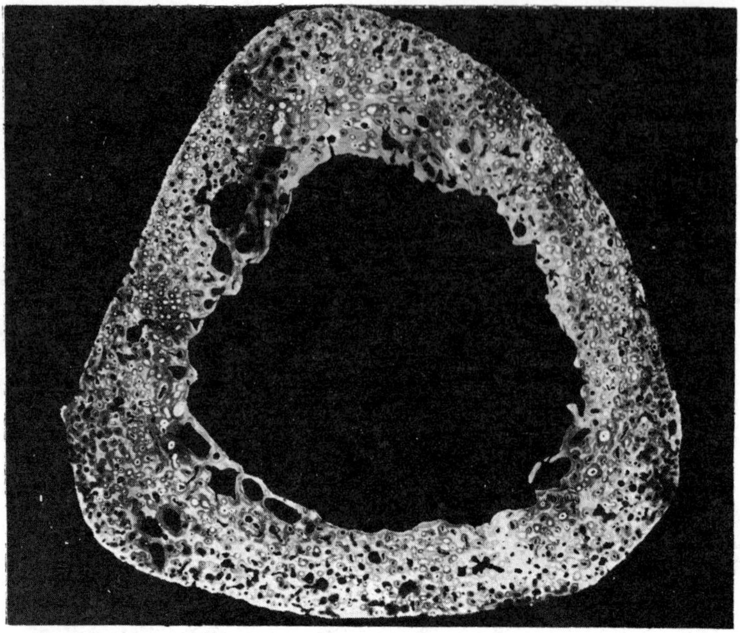

Figure 5. Photograph of a section of bone from the body of a former radium watch-dial painter. The dark areas are damaged bone.

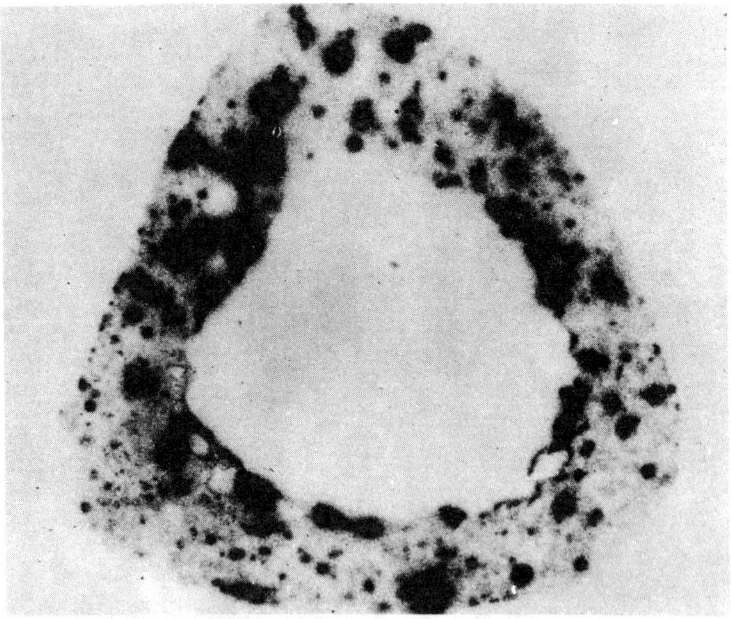

Figure 6. Autoradiograph of the bone section. In this picture the alpha radiation from the radium in the bone causes the darkening of the photographic plate.

The Test-Ban Treaty of 1963 was the result of the public outrage against the fallout of radioactive materials from atomic bomb tests. It would be a mistake to believe that everyone hailed the treaty:

"The fallout scare had one important practical consequence. It prepared the ground for the nuclear test moratorium. . . . Thus from insignificant and doubtful medical considerations, there seemed to follow a political and military measure of the greatest importance, which has contributed decisively to our weakness and our danger." [*The Legacy of Hiroshima,* by Edward Teller with Allen Brown (Doubleday and Co., New York, 1962)]

In the booklet *Fallout* (U.S. Atomic Energy Commission, 1963) C. L. Comar writes:

> The consensus of informed individuals is that the present or anticipated levels of radiation exposure from fallout due to nuclear tests (through 1962) do not constitute a hazard that warrants anxiety.
>
> There is also general agreement that fallout from nuclear testing contributes in a small way to worldwide radiation exposure and that the cost to society as a whole is probably not zero.

One of the people who drew attention to the possible dangers from fallout was Linus Pauling. The news of his receipt of the Nobel Prize for Peace was greeted by *Life* Magazine with an editorial entitled "A Weird Insult From Norway," (October 25, 1963), and by the *Bulletin of Atomic Scientists* with "A Welcome Honor from Norway" (December, 1963).

Nuclear Power

In 1975 there were 55 nuclear power plants in the U.S., generating about 9% of our electricity. A map showing the completed and planned reactors at that time is shown in Fig. 7.

The question of how much environmental radiation a reactor produces during normal operation was addressed in a series of advertisements which appeared in a number of national magazines including *Time* and *Newsweek*) in 1971. An example is shown in Fig. 8.

Figure 9 shows another advertisement which appeared in 1972.

One of the best articles by a proponent of nuclear power is the article by Hans Bethe in the January 1976 *Scientific American.* It is titled "The Necessity of Fission Power," subheaded with the statement "If the U.S. must have sources of energy other than fossil fuels, the only source that can make a major contribution between now and the end of the century is nuclear fission."

Among the greatest enthusiasts for unlimited nuclear power is Alvin M. Weinberg. A paper by him and R. P. Hammond was presented at a conference in 1971 and is reprinted in the *Bulletin of the Atomic Scientists* for March 1972. It begins with this plaintive statement:

We in the peaceful nuclear energy community have been com-

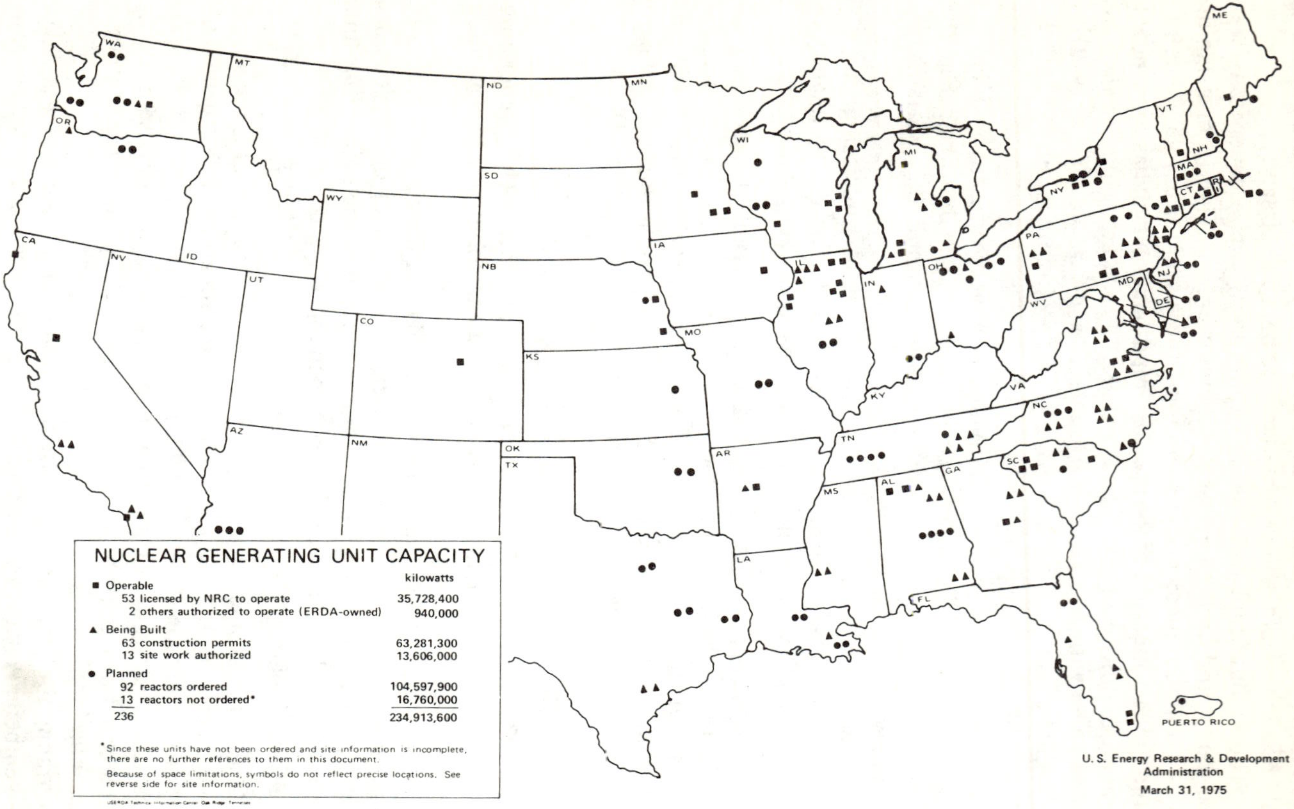

Figure 7. Nuclear Power Reactors in the United States

Should Denver be evacuated?

The higher the altitude of a city, the greater is its exposure to cosmic radiation. The people who live in mile-high Denver, for example, get more than three times the amount of radiation as those who live at sea level simply because there's less air over Denver to filter the sun's rays. Interestingly, in Denver, the average cosmic ray exposure alone is greater than the total amount of radiation the average person gets in New Jersey from all natural sources.

This comparison is made to put in clear perspective the concern some people might have about nuclear energy plants near their community. Anyone who worries about the one or two millirems* of radiation per year he may be exposed to from neighboring nuclear plants will probably be surprised to learn that over a hundred times more radiation from natural sources is all around him. Therefore, if one is really interested in his total radiation exposure, he can't ignore where he lives or works and what he eats.

For example, the radiation in all the food and water you eat and drink amounts to about 25 millirems a year. Is your home wood, stone, or brick? Wood gives off 50 millerems of radiation a year, concrete about 70, and brick as high as 100.

Every time you take a jet flight, you add a few more millirems to your body. Watch TV and add still a few more. From just one chest x-ray you're exposed to 50 millirems. A complete dental x-ray gives you 5,000 millirems, and a whole body diagnostic x-ray as much as 100,000 millirems.

Based on these facts, it should be clear that the radioactivity released from nuclear plants (under strict federal, state, and local regulations) is not a new kind of exposure. And the amount is only a tiny addition to the broad spectrum of natural background radiation and man-made radiation we all receive in our daily lives. It really isn't cause for worry.

Neither is the cosmic radiation level in Denver.

Salem Nuclear Generating Station. This plant is well under construction and is scheduled to start operation in 1974. It will have two units, each with a capacity of more than one million kilowatts. Another PSE&G plant — the Newbold Island Nuclear Generating Station — is scheduled to begin operating in 1978. Both plants will meet our ever-growing demands for electric power while preserving the natural resources and beauty of our land.

*an accurate unit of measure of biological effect on man

PSE&G

The Energy People are Environment People, too

Figure 8.

Mom's apple pie is radioactive. So is Mom.

That doesn't make her a dangerous woman.

Radiation is just naturally everywhere. In the earth. In your homes. In your food. In your Mom.

And each accounts for more radiation than a nuclear-power plant.

Take your Mom. Or yourself for that matter. The radioactive materials in our bodies give us an internal exposure of natural radiation averaging 25 millirems. (A millirem is 1/1000 of a rem, the standard unit of measurement of the biological effect of radiation.)

Buildings and the earth add another 55 millirems. Cosmic rays from space expose us to around 40 millirems. This varies—the higher the elevation you are at, the higher the exposure. All exposures considered, each American averages about 125 millirems of natural radiation annually.

Now, how about a nuclear power plant? Present operating experience tells us this: a person living anywhere in the vicinity of a typical nuclear power plant, 24 hours a day for a full year, would be exposed to less than 5 millirems of radiation from the plant.

Less than 5 millirems. Why, Mom could easily be exposed to that much during one round-trip coast-to-coast airline flight at 35,000 feet.

Those are the facts. And we think everyone should be aware of them. Because America's need for electricity is expected to double in the next ten years. To meet this need in an orderly fashion, clean, safe nuclear power must play an increasingly important role.

Our country's ability to do the work that needs to be done will depend on an adequate supply of electricity. There's no time to waste. New generating facilities must be built, and built in a way compatible with our environment.

We'll continue working to do this. But we need your understanding today to meet tomorrow's needs.

The people at your Investor-Owned Electric Light and Power Companies.

*For names of sponsoring companies, write to Power Companies 1345 Avenue of the Americas, New York, New York 10019

Figure 9.

> fortable in the belief that what we have wrought over the past 30 years has been an unmitigated blessing for mankind. It comes as a disconcerting shock therefore to find that, just when nuclear energy is clean, safe, and necessary, critical voices, particularly in the United States, claim it is unclean, unsafe, unnecessary.

The paper concludes that "mankind must have an alternative, essentially inexhaustible energy source. From what we now know this source must be nuclear," and "there probably are no insuperable global effects even if nuclear energy from fission breeders reaches 60 times the total energy man now produces."

Not everyone is equally enthustiastic: Lorna Salzman, writing for the Friends of the Earth, writes to the *Bulletin* (November 1975): "There is only one politically, biologically, and ethically acceptable solution: Total permanent abandonment of nuclear power." Gofman, writing in the journal *Environmental Action* for November 25, 1972, has called for "an immediate moratorium on all nuclear fission power generation."

Not everyone is so light-hearted as Richard Wilson, writing in a letter to the *Bulletin of the Atomic Scientists* (November 1975): "Attempting to ensure that there is no excessive hazard in a nuclear reactor accident is somewhat akin to ensuring there is no danger from dinosaurs when crossing the Brazilian jungle."

Up to this point we have not considered the economics of power generation. Although it is outside the scope of our present discussion, cost is clearly a decisive element in the choice among the different technologies. The discussion has implied that if the various dangers of nuclear reactor operation could be held in check, nuclear energy would be very desirable. This point of view has been challenged by Barry Commoner. He believes that we have sufficient oil and coal to make nuclear energy unnecessary and that, moreover, the energy from these fuels will be cheaper. He first describes the basis for his conclusion that slow-neutron reactors are too expensive, and then continues:

> Thus the need for the breeder reactor, a far more complex machine than the conventional one, has been removed by a kind of technological irony: The conventional reactors have themselves become so much more complex, and therefore expensive, that the reduced cost of fuel which the breeder was supposed to achieve is much less economically significant. The cost of electricity now depends much more on capital costs than on fuel costs.

Commoner concludes that

> what emerges from these considerations is the likelihood that the entire nuclear program is headed for extinction. It will leave us with a monument, which people will need to care for with vigilance, if not affection, for thousands of years: huge stores of radioactive wastes and the powerless, radioactive hulks of the

reactors that produced them. [*The Poverty of Power* (A. Knopf, New York, 1976)]

The quotations illustrate the wide variety of points of view about nuclear power. Many people consider the continued development of nuclear power to be essential; others believe it to be disastrous.

In order to facilitate a rational assessment, some of the questions which need to be considered are listed in Table IV. Some general remarks and brief comments on the table follow.

Table IV. Some Factors to be Taken into Acount in Considering the Development of Nuclear Reactors

Economic factors
capital cost
fuel cost
fuel reprocessing
Environmental problems
- mining
- atmostpheric radioactivity
- thermal pollution
- waste disposal

Accident hazard
- normal reactor operation
- transportation accidents
- catastrophic accidents
- insurance

Political questions
- fuel availability
- fuel diversion for bombs
- sabotage

Some of the Options

All present commercial reactors are of the slow-neutron type. Their fuel is ^{235}U, which is less than one percent of natural uranium. In terms of the energy to be recovered, the earth's resources of ^{235}U are of the same order or magnitude as the remaining reserves of coal.

To look further ahead is risky. We need only to imagine ourselves a hundred years ago trying to foresee the technical developments of the century that lay ahead.

Given enough time we can probably develop means of using energy for which the raw materials are abundant, or even limitless. The breeder reactor has to be considered as one such possibility. It converts the abundant uranium isotope ^{238}U into plutonium, which can be used as a reactor fuel.

Perhaps the fusion process, by which the sun produces the energy which it radiates, can be used. So far this has not been demonstrated. If and when a fusion reactor becomes available it is certain that the problems which will accompany its use will not be minor.

A particularly attractive possibility is that the most benign source of all, solar energy, will be used. We know how to use it, but not very well. The "fuel" costs nothing, but the difficulty and cost of converting it to usable heat or electricity have until now been prohibitive.

It seems that for the next decade at least, and perhaps for longer, we are stuck with the fossil fuels, of which coal is the most abundant, and with nuclear reactors. The comments which follow deal only with the slow-neutron reactor. Brief comparisons with the breeder and with coal as a fuel conclude this section.

Economic Factors

The cost of nuclear reactors for power generation is obviously competitive, or none would be built. At the same time it may be useful to remember that the true cost to the citizen may be difficult to determine, since it depends heavily on government subsidies and on the tax structure. It becomes particularly hard to assess if we wish to include as "cost" the degradation of the environment.

Fuel Reprocessing

A step in the fuel cycle which has been widely discussed, but not yet exploited, is fuel reprocessing. This is the recovery of the ^{235}U which remains, and the separation of the plutonium which has accumulated.

The manufacture of plutonium, which is a principal process in the breeder, takes place to some extent in all reactors. Unfortunately plutonium is a particularly dangerous material. The inhalation of a fraction of a microgram is lethal.

There is another danger: a few kilograms of plutonium are enough to make a fission bomb. This is true of ^{235}U also, but ^{235}U is never used pure in a reactor, and is difficult to separate from natural uranium. Even "enriched" uranium, which is often used as reactor fuel, and in which the concentration of ^{235}U is several percent, cannot be used to make bombs.

Atmospheric Radioactivity

Among the products of fission are the noble gases krypton and xenon.

These do not combine with other elements; they are released into the atmosphere. Some tritium is also produced and released into the environment. The estimated amount of radioactivity is shown on Figure 1, and forms the subject of the advertisements reprinted here as Figures 8 and 9.

The radiation hazard is seen to be small compared to the normal background. The total amount of radioactive material released into the air is, however, huge, and may have unexpected consequences. A critical isotope is krypton 85 with a half-life of about eleven years. The ionization produced by this isotope may become so large that it will affect the electrical properties of the atmosphere, and hence rain formation and thunderstorms.

An article called "Meteorological Consequences of Atmospheric Krypton-85" (*Science* **193**, 195, 1976) by W. L. Boeck describes the problem and concludes: "If release of ^{85}Kr into the atmosphere continues unabated, global changes in the atmospheric electric circuit will occur within 50 years. Our present understanding of atmospheric processes is insufficient to determine the extent of consequent weather changes and whether they would be beneficial or harmful."

Thermal Pollution

In a typical nuclear power plant about 30% of the fission energy is changed to electrical energy. That means that for every kilowatt-hour of electrical energy more than twice as much heat is produced and must be absorbed by the environment. Usually a river is used, or the ocean. It is also possible to construct cooling towers which allow the heat to be dissipated in the atmosphere.

The fish population is drastically affected by changes in water temperatures. Different species are attracted to the vicinity of a power plant; unfortunately, they may not survive the inevitable shutdowns.

Waste Disposal

During the operation of a nuclear reactor the fuel, uranium or plutonium, is converted to fission products. These include a wide range of elements near the middle of the periodic table. The problem is that the nuclides which are produced are highly radioactive. Some decay very quickly, whereas others have long half-lives. Important nuclides with relatively long half-lives are strontium 90 (28 years), cesium 137 (30 years), and krypton 85 (11 years). These are not, however, the only radioactive materials which are produced. Neutron bombardment also leads to radioactive isotopes of the heaviest elements. The most important is plutonium 239, whose half-life is 24,000 years.

The radioactivity of the reaction products is so great that it produces a large amount of heat. If a nuclear reactor has been operating for some months and its chain reaction is stopped, it will continue to produce heat. The amount of power immediately after shutdown will be 7% of the amount before shutdown. After one day the heat generated by the fission products drops by a factor of 13, after one year by a factor of 340, and after ten years by a factor of 3,400.

Although their radioactivity is enormous, the volume of the fission products is not very great. Present power reactors produce 500 to 1,000 megawatts of electricity. A 1,000 megawatt (10^9 watt) reactor uses about a ton of ^{235}U in a year. In its pure form this would represent a volume of about a cubic foot. This is therefore also the order of magnitude of the volume of the pure solid fission products which are produced. In addition, however, there are much larger volumes of low-level wastes in which the concentration of radioactive materials is relatively small.

Proposals for dealing with the radioactive waste include storing it in salt mines and burying it in the Arctic ice. No completely satisfactory long-term method has so far been adopted. Millions of curies have been poured into the earth and into the ocean. Now the fission products are largely kept as liquids in steel tanks where they continue to boil for long times because of the heat which they generate.

A good deal has been written about the necessity of guarding the fission products for thousands of years. Plutonium is clearly the biggest long-term problem and should be considered separately. If the plutonium, the ^{137}Cs and the ^{90}Sr are separated, the rest of the fission products will become relatively inert in a time measured in decades.

Accidents

The problems which we have discussed so far are technical and can be analyzed more or less straightforwardly. Their solutions may be difficult and costly but they are nevertheless predictable. There are other problems which are inherently more unpredictable, and therefore particularly vexing.

Although there have been no accidents in the nuclear power industry which have led to fatalities from radiation (but see problem #5, in Problem Set 2), there have been, in addition to the Three Mile Island accident, surprisingly many minor failures in reactor operation. Most have been as ordinary as plumbing leaks, which can, of course, be very dangerous when they involve highly radioactive materials. If fuel reprocessing becomes common, with frequent transport of fuel and fission products, the opportunities for accidents will increase further.

A good deal of attention has been paid to the possibility of major accidents. There are two questions: how likely is such an accident, and how bad can it be? A study made for the American Physical Society concludes: "We have not found a basis for substantial short-range concern about the safety of light-water reactors, nor, on the other hand, have we found a completely satisfactory quantitative treatment of the important safety issues." It recommends "a greatly expanded . . . program in reactor safety research." (Light-water reactors are slow-neutron reactors using ordinary water as a moderator. All present American power reactors are of this type.)

A reactor safety study made for the government (the "Rasmussen Report") attempted to calculate the likelihood of a major accident, but the American Physical Society study comments that "based on our experience with problems of this nature involving very low probabilities, we do not now have confidence in the calculated absolute values of the probabilities."

The general conclusion seems to be that an accident which results in several thousand deaths is indeed conceivable, but that it is so unlikely that the average risk per person is much less than other industrial risks which we accept.

Political Questions

The specter of a terrorist group with a crude atomic bomb made from stolen plutonium is a frightening one. On the other hand it has been argued that the theft and the bomb manufacture demand considerable sophistication, and that other means for destruction and blackmail exist and may be more accessible.

Similarly, the possibility of sabotage of a reactor must be considered. Whether a reactor is a more likely or more dangerous target than some other industrial installation is open to question.

A final political question is that of fuel availability. We are now heavily dependent on oil imports, particularly from the Middle East. Some of the oil is used for the generation of electricity and this fraction could be eliminated by the substitution of other fuels, or of nuclear power.

The Breeder Reactor

The breeder reactor has the advantage that it can free us from the problem of fuel supply. Since it runs at a higher temperature it is more efficient than a water-cooled reactor.

Because of the higher temperature at which it operates, the breeder cannot be cooled by water. Present designs call for cooling with liquid sodium or high-pressure gas. The breeder is therefore substantially more complex and hazardous. The problems accompanying the use of plutonium (toxicity and theft hazard) would become much more important with the widespread use of breeders. The resulting "plutonium economy" is considered intolerably dangerous by many people.

Coal

Because of its familiarity we tend to forget what an unpleasant and dangerous fuel coal is. Deep mining of coal is among the most hazardous occupations. Strip mining devastates the land. The burning of coal causes enormous amounts of chemicals and small particles to be emitted into the atmosphere. One attempt to estimate the risk to life from coal-burning power plants comes to the conclusion that it is about 100 times as great as from nuclear plants ["Health Effects of Energy Production and Conversion," by C. L. Comar and L. Sagan in *Annual Review of Energy*, Vol. 1, 1976].

Epilogue

The production of usable energy from fossil fuels and from nuclear reactors is associated with substantial detrimental effects on our health and environment. An awareness of the dangers is now slowly causing increased interest in more benign technologies and in ways to use energy more efficiently and to husband our resources more jealously. It is to this informed awareness that this booklet is dedicated.

Bibliography

The amount of written material which exists on radiation and its effects is enormous. Here are a few references, some of which have been mentioned before, which are particularly appropriate:

A short review with an emphasis similar to that which we have used is the chapter called "Radiation" by J. Harte, R. H. Socolow, and J. N. Ginocchio in the book *Patient Earth* by Harte and Socolow (Holt, Rinehart and Winston, 1971).

An excellent short discussion of the genetic effects is in the booklet by I. Asimov and T. Dobzhansky referred to earlier [*The Genetic Effects of Radiation,* "Understanding the Atom" series (U.S. Atomic Energy Commission, Division of Technical Information, 1966)].

A more complete discussion appears in *Radiation, Genes, and Man*, by B. Wallace and T. Dobzhansky (Holt, Rinehart and Winston, 1959).

A comprehensive popular book is *Radiation, What It Is and How It Affects You* by J. Schubert and R. E. Lapp (Viking Press, 1957).

An article which emphasizes and describes the difficulty of estimating cancer risks is Charles E. Land's "Estimating Cancer Risks From Doses of Radiation," *Science*, Vol. 209, p. 1197 (1980).

The book *Nuclear Power and the Public*, edited by H. Foreman (University of Minnesota Press, 1970), contains a number of papers with widely different points of view, and includes the audience's questions, from a public symposium. There is a good deal of discussion about radiation levels and the resulting health hazards, but, curiously, the possibility of accidents is barely mentioned.

It need hardly be reiterated that the BEIR reports (published in 1972 and 1980) are essential references [*The Effects on Population of Exposure to Low Levels of Ionizing Radiation* (National Academy of Sciences, National Research Council, Washington, DC 20006)].

There is an excellent, comprehensive annotated bibliography which covers the time to just before BEIR I was issued. It is essential for all who are interested in any part of the subject: "Resource Letter ERPEE-1, Energy: Resources, Production, and Environmental Effects," by R. H. Romer, *American Journal of Physics,* Vol 40, p. 805 (June, 1972).

Very useful material may be found in the *Annual Reviews of Energy* (Annual Reviews, Inc., Palo Alto, California 94306). Volume I (1976), for instance, contains 28 papers. Of particular relevance is an article on the nuclear fuel cycle and a group of five articles on environmental costs, health, safety, and risk analysis.

The "Report to the American Physical Society by the Study Group on Light-Water Reactor Safety" was published as Supplement No. 1 (Summer 1975) to volume 47 of the *Reviews of Modern Physics.* It contains detailed descriptions of the two most important types of American reactors, and contains a large amount of information on reactors. There are, for example, lists of fission products with their half-lives, the amounts to be expected, the heat which they produce, and their possible effects in case of an accidental release.

Projects and Discussion Questions

1. Discuss the advertisements shown in Figs. 8 and 9. Is the scientific information correct? What are the results of the extra radiation in the light of the answers to problems 1 and 2 of Problem set 2? Are the results negligible? What is the emotional impact of the ads?
2. Discuss the correctness and the limitations of the statement on the innocuousness of x rays on p. 23.
3. List and discuss the advantages and disadvantages of nuclear reactors for electric power generation.
4. Discuss your recommendations concerning the continued development of nuclear reactors for electric power generation.
5. Read and summarize the *Report of the President's Commission on the Accident at Three Mile Island* (U.S. Government Printing Office, Washington, DC, October 1979.)
6. Choose one of the books or articles in the bibliography. Read it, summarize it, and discuss some of the results or implications.

5. *Experiments*

Two kinds of simple physical experiments are particularly appropriate for this sequence. One illustrates the absorption characteristics of radioactive radiations, the other measures the rate of decay.

The details of the experimental procedure depend on the particular equipment which is used and are usually described in the manufacturer's literature.

Apparatus Notes

For absorption experiments it is necessary to have radioactive sources, absorbers of different materials and thickness, and a detector. Most of the companies which manufacture educational equipment sell complete setups for this purpose. Examples are:

Cenco No. 71201-2 radioactivity demonstrator

Ealing No. 32-2669 ratemeter (sources sold separately)

Klinger No. KA 6433 radioactivity teaching system

Sargent-Welch No. 2192 radioactivity experiment system and demonstrator

Lab manuals which describe the use of the equipment and the suggested experiments are usually included.

Half-life experiments are best made with short-lived daughter products which may be separated from a long-lived parent material. A particularly simple apparatus is made by Sargent-Welch (No. 0619 emanation electroscope). It uses thorium oxide as the parent from which radon 220, whose half-life is less than a minute, may be separated by squeezing a rubber bulb. The radioactive material remains in the apparatus, and no chemical procedures are required.

Other parent-daughter combinations are made by the Union Carbide Corporation and sold through most apparatus companies under the name "minigenerators." The short-lived daughter substance is separated as a solution so that the procedure is somewhat less foolproof than with the thorium-radon source. For a small class having adequate supervision those sources are, however, more versatile. A lab manual is available.

Appendix

A Summary of Some Properties of Nuclei, Nuclear Radiation, and Reactors

Peter Lindenfeld
Rutgers University

1980
Revised 1982

Preface

This summary is intended as background for those Issue-Oriented Modules which are based on nuclear and radiation physics. It reviews and supplements some of the material which is usually found in elementary physics texts.

Prerequisites

Study of this summary and its associated modules should accompany or follow an elementary physics course. The student is expected to be familiar with the law of conservation of energy, mass-energy equivalence, and elementary notions of atoms such as the Bohr model, Avogadro's number, and the microscopic, kinetic interpretation of heat energy.

Other prerequisites are: the basic units of the SI (MKS) system, conversion of units, scientific notation (powers of ten), and significant figures. For the sections on the radioactive decay equation and half-life, a knowledge of logarithms and exponentials is required.

Appendix: Contents

1. Composition of Atoms and Nuclei

An atom consists of a *nucleus* and *electrons.* The nucleus is positively charged; the electrons are negatively charged. The electrical force between them holds the electrons to the nucleus.

All electric charge is quantized in units of the electronic charge. We will use the symbol e for the magnitude of that charge, which is 1.6×10^{-19} coulombs. The charge on an electron is then $-e$.

A nucleus consists of *nucleons.* There are two kinds of nucleons: *protons* and *neutrons.* The proton has a charge of $+e$; the neutron has no charge.

The *periodic table of elements* is arranged in order of the *atomic number,* Z. This is the number of protons in the nucleus. In a neutral atom Z is also the number of electrons which surround the nucleus. An atom may, however, have more or less than Z electrons; when this is the case it is called an *ion.*

An element may be specified by giving its name or by giving its atomic number. The eighth element in the periodic table, for example, is oxygen. Its atomic number is $Z=8$, and there are eight protons in the nucleus of each oxygen atom. A neutral atom of oxygen has eight electrons which surround the nucleus.

The mass of a nucleon is about 1840 times that of an electron. Therefore almost all of the mass of an atom is that of its nucleus.

The nucleus, on the other hand, takes up only a tiny fraction of the *space* occupied by an atom. The diameter of an atom is about four orders of magnitude (10^4) times as great as that of its nucleus. Most of the volume of the atom consists of empty space.

The number of neutrons in a nucleus is the *neutron number,* N. The number of nucleons in a nucleus is the *mass number,* A, so that $A=Z+N$.

A particular kind of nucleus, or *nuclide,* is specified by the notation ${}^{A}_{Z}X$ where X is the symbol for the particular element whose atomic number is Z. (This notation is a recent change. The previous notation was ${}_{Z}X^{A}$.) The symbol ${}^{17}_{8}O$ denotes a nucleus of oxygen (O), atomic number $Z=8$, with a total number of nucleons $A=17$. The number of neutrons is $A-Z=9$. Since oxygen is element number 8, the number for Z may be omitted, and we can write simply ${}^{17}O$.

Nuclides with the same Z but different N are called *isotopes.* Since they have the same Z they belong to the same element and have the same chemical properties. Ordinary oxygen, for example, consists of a mixture of three isotopes: ${}^{16}O$, ${}^{17}O$, and ${}^{18}O$. (The term "isotope" is also commonly used as a synonym for "nuclide," but we will avoid this usage here.)

The nuclei of the isotopes of hydrogen have special names. "Ordinary" hydrogen comprises 99.985 percent of natural hydrogen, with $A=1$. Its nuclei are protons. The rest (0.015 percent) has $A=2$ and is called *deuterium.* The nucleus of a deuterium atom is a *deuteron.* There is also a radioactive isotope with A=3. It is called *tritium* and its nucleus is a *triton.* The symbol D is sometimes used for ^{2}H and the symbol T for ^{3}H.

Nuclides with the same A but different Z and N are called *isobars.* $^{17}_{8}$O and $^{17}_{9}$F are isobars.

2. Mass and its Equivalent Energy

Tables of "nuclear" masses are normally not the masses of nuclei, but rather those of the neutral atoms. It is possible (and often even desirable) to work with the atomic masses. However, if the actual nuclear mass must be known, the mass of Z electrons must be subtracted from the entry in the tables.

The mass unit for atomic particles is the *unified atomic mass unit,* with the symbol u. One u is equal to one-twelfth the mass of a neutral ^{12}C atom. One u is equal to 1.66×10^{-27} kg.

Einstein's relationship $E=mc^2$ describes the equivalence of mass and energy. Since $c=3 \times 10^8$ m/s, it shows that one kilogram is equivalent to 9×10^{16}J. All mass units and energy units can similarly be converted into one another. A common energy unit is the *electron volt* (eV): 1 eV$=1.6 \times 10^{-19}$J and 1 MeV$=1.6 \times 10^{-13}$ J. It follows (if we use the required number of significant figures) that 1 u is equivalent to 931.5 MeV. This allows us to say, for example, that the mass of a proton (1.007 u) is equivalent to an energy of 937.5 MeV, or simply (as a kind of shorthand way of saying the same thing) that the mass of the proton *is* 937.5 MeV. (Table A-2 gives the masses of the electron, proton, and neutron in various units.)

In relations which include masses and energies, the conversion factor, c^2, is often not included explicitly. As always, each term in an equation must have the same units. These may be mass units, such as kg or u, or energy units, such a J or MeV.

3. Decay Constant, Half-life, Activity

Radioactive decay takes place in accordance with the equation

$$N=N_o e^{-\lambda t} \tag{1}$$

where N is the number of nuclei present at time t, and N_o is the initial number of nuclei at the time t=0. The time for N to drop to N/2 is called the *half-life,* $T_{1/2}$. λ is called the *decay constant* and its reciprocal the *mean life.* The product λt is dimensionless, so that λ has units of s^{-1}.

The number of disintegrations per unit time is equal to λN, and is called the *activity.* The symbol A is sometimes used for activity. The SI unit for activity is the becquerel (Bq), equal to one disintegration/sec. An older, but still widely used, unit is the curie (Ci) equal to 3.7×10^{10} Bq. It was originally defined to be the activity of one gram of radium.

The activity per unit mass is called the *specific activity*. It may be measured in Bq/kg or in Ci/g.

As an example consider a radioactive material whose half-life is one year. After one year the original number of nuclei, N_o, will have dropped to $N_o/2$; after two years to $N_o/4$; and after three years to $N_o/8$. In general, after a time equal to $nT_{1/2}$ the amount of material has dropped to 2^{-n} times the original amount. (The amount of material can be specified by giving the number of atoms which are present, or by giving its mass.)

Equation 1 can also be written $A=\lambda N=\lambda N_o e^{-\lambda t}$. This form shows that the activity decays according to the same exponential law, with the same decay constant and the same half-life.

Several other forms of the equation are useful:

$$\frac{N}{N_o}=e^{-\lambda t}, \quad \frac{N_o}{N}=e^{\lambda t},$$

or by taking natural logarithms of both sides:

$$\ln\frac{N_o}{N}=\lambda t \tag{2}$$

When $N = N_0/2$, the time t is $T_{1/2}$. If we substitute these values in Eq. 2 we see that $\ln 2 = \lambda T_{1/2}$. ($\ln 2 = 0.693$)

It is often important to calculate the number of atoms (N) in a specimen of a given mass. It is then necessary to remember that there are 6.023×10^{23} (Avogadro's number) atoms in one mole of material. It is also useful to remember that to an approximation of better than one percent the number of grams in a mole of material is equal to the mass number of its nucleus. (The exact atomic mass is different, and depends on the binding energy. The binding energy is discussed later.) In other words, there are about 6×10^{23} atoms in 16 g of ^{16}O, as well as in 226 g of ^{226}Ra.

Example:

How long does it take for 1 gram of ^{131}I to decay by a factor of 8, and by a factor of 100? Its half-life is 8.05 days.

(a) $8=2^3$; hence a radioactive material decays by a factor of 8 in 3 half lives, or, in this case, in 24.15 days.

(b) If the time is not equal to an integral number of half lives we have to use the decay equation $N=N_o e^{-\lambda t}$. Here $N_o/N=100$ so that $\lambda t=\ln 100$ which is equal to 4.605. (This is the logarithm to the base e, or the natural logarithm.)

The time is therefore $t=4.605/\lambda$, or, since $\lambda=0.693/T_{1/2}$ it is $(4.605)(T_{1/2})/.693$, or $6.65\ T_{1/2}$, which is 53.5 days.

Example:

The radioactive nuclide ^{40}K has a half-life of 1.3×10^9 y. What is its specific activity in Ci/g?

$$\lambda = \frac{0.693}{T_{1/2}} = \frac{0.693}{1.3 \times 10^9\,y}\ \frac{1\,y}{3.16 \times 10^7\,sec}$$

Let N be the number of atoms in 1 gram. Since 40 g contain 6×10^{23} atoms, $N = 6 \times 10^{23}/40$ atoms/g.

The specific activity in Bq/g is

$$\lambda N = \frac{0.693}{1.3 \times 10^9}\ \frac{1}{3.16 \times 10^7}\ \frac{6 \times 10^{23}}{40}\,Bq/g$$

and, since 1 Ci $= 3.7 \times 10^{10}$ Bq it is also

$$\lambda N = \frac{0.693}{1.3 \times 10^9}\ \frac{1}{3.16 \times 10^7}\ \frac{6 \times 10^{23}}{40}\ \frac{Bq}{g}\ \frac{1\,Ci}{3.7 \times 10^{10}\,Bq}$$

$$= 6.9 \times 10^{-6}\,Ci/g \text{ or } 6.9\,\mu Ci/g.$$

Problem Set A-1

1. The half-life of ^{85}Kr is 11 y. 160 Ci of ^{85}Kr are released from a nuclear reactor. How many years does it take for 150 Ci of this amount to decay?
2. (a) How many half-lives does it take for a radioactive specimen to decay to 1/10 percent (.001) of its original amount? Use your result to two significant figures to calculate the following:
   (b) How many half-lives does it take for 1 Ci to decay to 1 μCi?
   (c) How many atoms are left after 70 half lives if the original amount is 1 gram-atomic mass?
3. (a) Show that the number of half-lives for a reduction by a factor of 10 is ln 10/ln 2.
   (b) Use your result from part a to find the number of half-lives for a reduction by factors of 10^3 and 10^6, and compare to the answers of problem 2, parts a and b.
4. The half-life of tritium, ^3_1H, is 12.3 years. What is its specific activity in Ci/g?
5. The half-life of ^{238}U is 4.5×10^9 y. What is the mass of 1 Ci of ^{238}U?
6. ^{90}Sr has a half-life of 28 years. How long will it take for 1 g of ^{90}Sr to decay to 1 Ci?

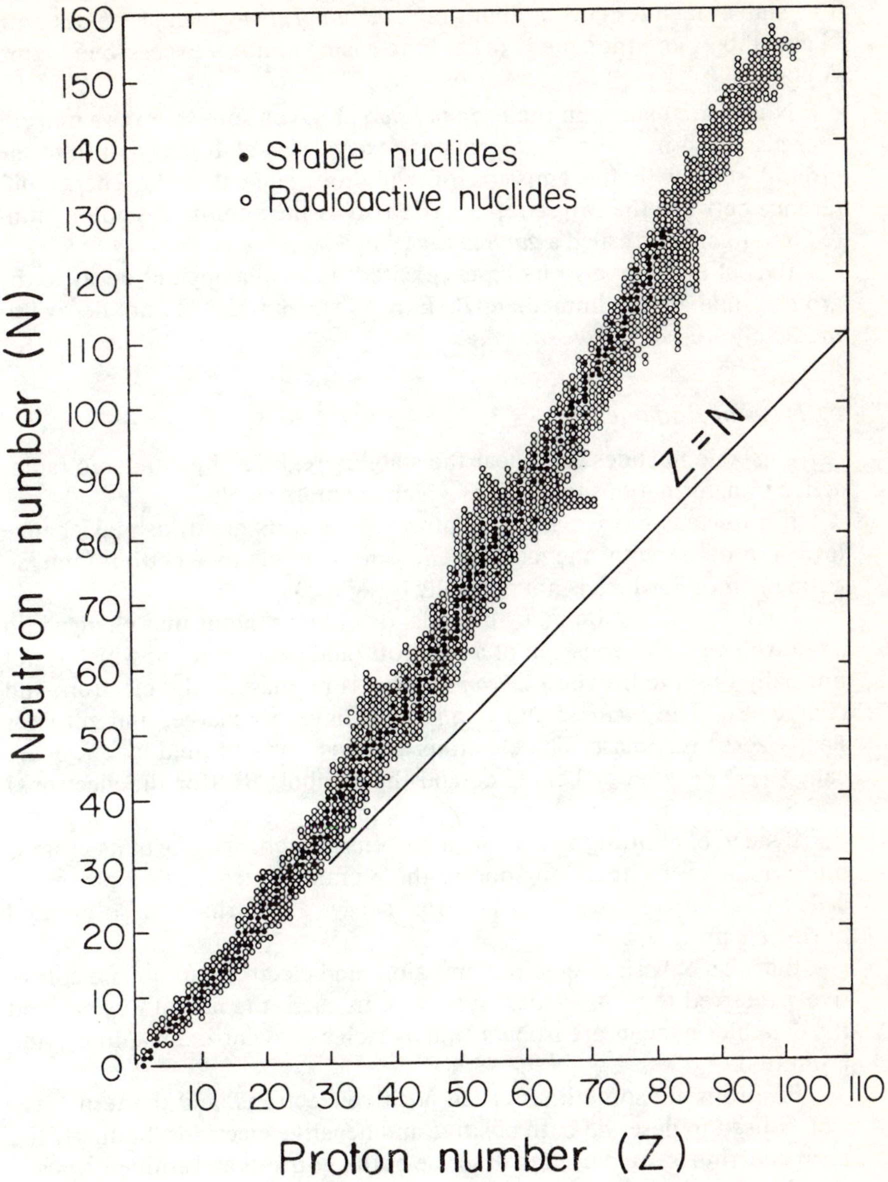
Neutron number (N)
Proton number (Z)
• Stable nuclides
○ Radioactive nuclides
Z=N
0
10
20
30
40
50
60
70
80
90
100
110
120
130
140
150
160
10
20
30
40
50
60
70
80
90
100
110

Figure A-1.

4. Stable Nuclides, Energy States, Gamma Rays

Each nuclide may be represented by a point on a graph of N against Z. The stable nuclides cluster about the *stability region* which starts out with N equal to Z, and then tends toward increasing neutron excess. See Figure A-1.

Nuclei are usually in their *ground state.* If given some energy a nucleus may also exist in higher energy states *(excited states).* It may return to the ground state with the emission of a photon, such that the energy difference between the two states is carried off by the photon. A photon emitted by a nucleus is called a *gamma ray* (γ ray).

Except in relatively rare cases, excited states of a nucleus decay to the ground state almost immediately. (Excited states which do not decay immediately are called *isomeric states.)*

5. Types of Radioactivity

Unstable nuclides exist near the stability region. They undergo radioactive transformations to nuclides which are more stable.

If a nucleus has too many neutrons, one of its neutrons may change into a proton, with the concurrent emission of an electron (and an antineutrino, $\bar{v}$, which is not normally observed).

If a nucleus has too many protons, one of its protons may change into a neutron with the emission of a positron (and a neutrino, v, which is not normally observed). The *positron* has the same mass as the electron, and charge +e. The *neutrino* and *antineutrino* have no charge, and zero (or nearly zero) rest mass. The electrons and positrons emitted by nuclei are called *beta particles* or beta rays, and the symbols β^- (for the electrons) and β^+ (for the positrons) are used for them.

Instead of emitting a positron, a nucleus may change one of its protons into a neutron by absorbing one of the atomic electrons. This process is called *electron capture.* As in positron decay, a neutrino is also emitted during electron capture.

Electron emission, positron emission, and electron capture are collectively referred to as *beta-decay processes.* In each the parent nucleus and the daughter nucleus are isobars, and in each a neutrino or an antineutrino is emitted.

Positrons are sometimes called positive electrons, and the term electron is used to describe both positive and negative electrons. Similarly the term neutrino is often used to describe both neutrinos and antineutrinos.

When positrons and electrons meet they undergo *annihilation.* They both disappear and their energy (including the rest energy) appears in the form of photons, which are called *annihilation radiation.* Most of the time annihilation takes place when the positrons and electrons are almost at rest, and results in the emission of two photons (in order to conserve momentum), each with an energy of 0.51 MeV, equal to the rest energy of an electron or a positron. Positron emission from a radioactive source is always accompanied by annihilation radiation from the surrounding material.

Some nuclei, especially those near the high-mass end of the stability region, disintegrate by emitting a ^{4}He nucleus. The emitted helium nucleus is called an *alpha particle* (α particle).

Radioactive decay processes may be represented by equations, such as the following:

electron emission: $^{60}_{27}Co \rightarrow ^{60}_{28}Ni + \beta^- + \bar{v}$

positron emission: $^{14}_{8}O \rightarrow ^{14}_{7}N + \beta^+ + v$

electron capture: $^{40}_{19}K + \beta^- \rightarrow ^{40}_{18}A + v$

alpha emission: $^{226}Ra \rightarrow ^{222}Rn + \alpha$

If the released energy is included explicitly, the decay equation may be used to describe the mass and energy transformations quantitatively:

$$^{121}Sn \rightarrow ^{121}Sb + \beta^- + \bar{v} + 0.38 \text{ MeV}$$

This expression can be read as follows: The rest energy of a nucleus of ^{121}Sn is equal to the rest energy of a nucleus of ^{121}Sb plus the rest energy of an electron, plus 0.38 MeV. (The neutrino has no rest energy.) The extra energy (in this case 0.38 MeV) appears as kinetic energy of the decay products. (Gamma rays are sometimes written explicitly in the equation. Sometimes their energy is included in the released energy.)

If the masses are such that there is no positive released energy on the right-hand side of the equation then the reaction cannot occur spontaneously.

The mass of the neutron is larger than the mass of the proton plus the mass of the electron. Therefore a free neutron can undergo beta decay, but a free proton cannot do so.

The relations of the decay equations may also be described by energy-level diagrams, or decay schemes. The energy levels of a nucleus are drawn one above the other; those of the nucleus with a higher Z are drawn further to the right. Here are some examples:

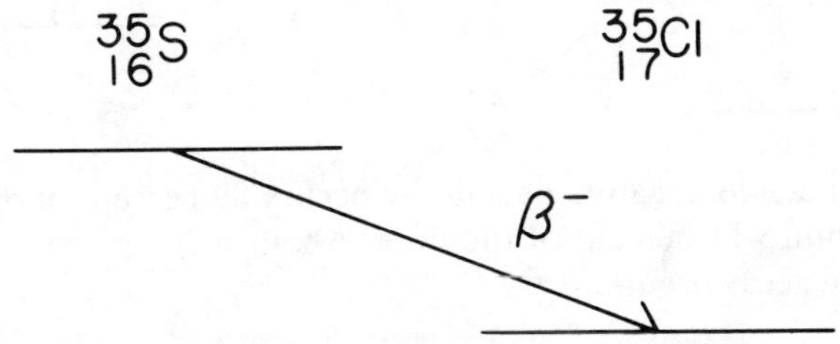

Each decay releases an energy of 0.17 MeV, equal to the difference between the initial and the final energy level. Some of this energy is given to the beta ray, some to the neutrino. The maximum energy which the beta rays can have is therefore 0.17 MeV.

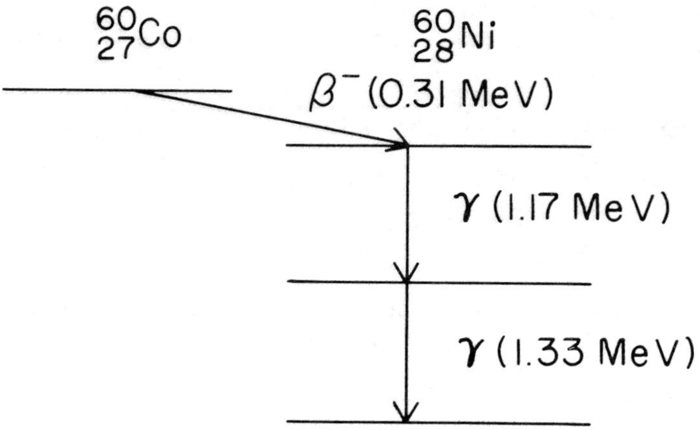

In this case the beta decay is to an excited state of ^{60}Ni, which immediately decays to another excited state and then to the ground state. Each beta ray is therefore accompanied by two gamma rays.

A nuclide which can undergo all three modes of beta decay is ^{40}K:

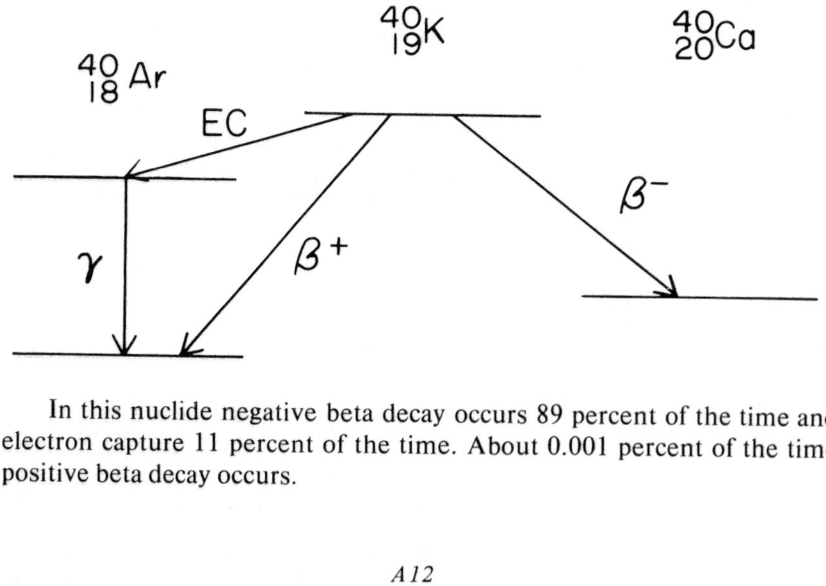

In this nuclide negative beta decay occurs 89 percent of the time and electron capture 11 percent of the time. About 0.001 percent of the time positive beta decay occurs.

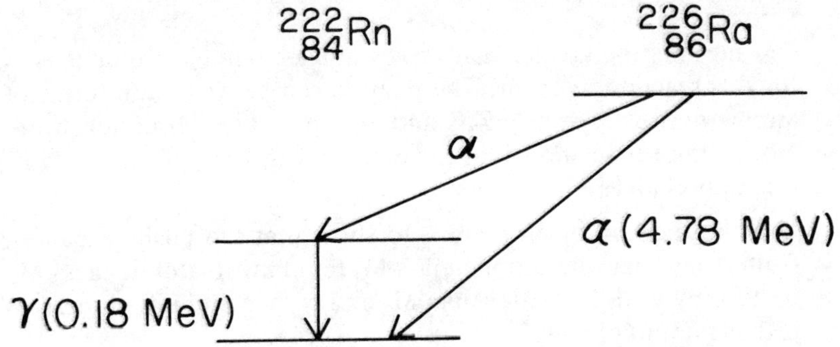

Ninety-five percent of the decays are directly to the ground state.

Radium is part of a natural radioactive "decay chain." It is the product of α decay from ^{230}Th, and its "daughter," ^{222}Rn, is also radioactive and emits α particles.

Problem Set A-2.

1. Write down the equations which describe the following radioactive transformations: (The symbol Q is used for the released energy, e.g.

$$^{35}_{16}S \rightarrow ^{35}_{17}Cl + \beta^- + \bar{\nu} + Q)$$

   (a) β decay of ^{90}Sr.
   (b) β decay of tritium.
   (c) α decay of ^{235}U.
   (d) the two possible modes of beta decay of $^{22}_{11}Na$ to $^{22}_{10}Ne$.

2. (a) Write down the equation for the decay of a neutron to a proton.
   (b) Use the masses of the neutron and the proton to calculate the maximum energy in MeV of the beta ray which is emitted in neutron decay.
   (c) What is the maximum energy of the neutrino which is emitted?
   (d) Why is it not possible for a free proton to decay into a neutron?

3. (a) Calculate the energy Q (in MeV) which is released in the reaction $^2H + ^3H \rightarrow ^4He + n + Q$ from the masses of the nuclei.
   (b) Table A-2 gives the masses of the neutral atoms. The masses of the nuclei are less by Zm_e. How many electron masses must be subtracted on the left-hand side of the reaction equation of part a. How many on the right-hand side?

4. A nucleus whose mass is m_1 and whose atomic number is Z undergoes negative beta decay to a nucleus whose mass is m_2 and whose atomic number is Z+1, with an energy release Q. The relation between the masses is therefore

$$m_1 = m_2 + m_e + Q$$

and m_1 must be greater than m_2 by at least one electronic mass (m_e) for this reaction to occur. The relation can be written in terms of the atomic masses $M_1 = m_1 + Zm_e$ and $M_2 = m_2 + (Z+1)m_e$. Substitute the atomic masses to show that the reaction can occur if (and only if) M_1 is greater than M_2.

5. Use the method of problem 4 to show that the positive beta decay from a material with atomic mass M_1 to one with atomic mass M_2 can occur only if M_1 is greater than M_2 by a certain amount of mass. What is that amount of mass?
6. A nucleus whose atomic number is Z can, in principle, decay to a nucleus whose atomic number is $(Z-1)$ either by β^+ decay or by electron capture. Use the method of problem 4 to show that it is sometimes possible for electron capture to occur when positron decay is not possible. What must the difference $M_1 - M_2$ between the initial and the final atomic masses be for positron decay to be possible? What must the difference be for electron capture to be possible?
7. ^{14}C decays to ^{14}N with an energy release of 0.16 MeV. Use the value of this energy and the atomic mass of ^{14}N in Table A-2 to calculate the atomic mass of ^{14}C.
8. Use the masses in Table A-2 to find the energy which is released when a nucleus of ^{235}U decays by α emission.
9. The energy-level diagram of ^{40}K shows that it can decay to an excited state of ^{40}Ar by electron capture, or to the ground state of ^{40}Ar by β^+ emission. The neutrino which accompanies electron capture has an energy of 0.05 MeV, and is followed by a gamma ray of 1.46 MeV. What is the maximum energy of the positron?

---

6. Binding Energy

The total binding energy, E_b, of a nucleus is the amount of energy which must be supplied to it to separate it into its A nucleons. The process can be written as

$$^{A}_{Z}X + E_b \rightarrow Z + Nm.$$

On the left-hand side of this expression are the original nucleus and the binding energy. On the right-hand side are the separated Z protons and N neutrons. In terms of the masses this relation shows that the binding energy is the energy equivalent to the mass of Z protons, plus the mass of N neutrons, minus the mass of the whole nucleus:

$$m_x + E_b = Zm_p + Nm_n$$

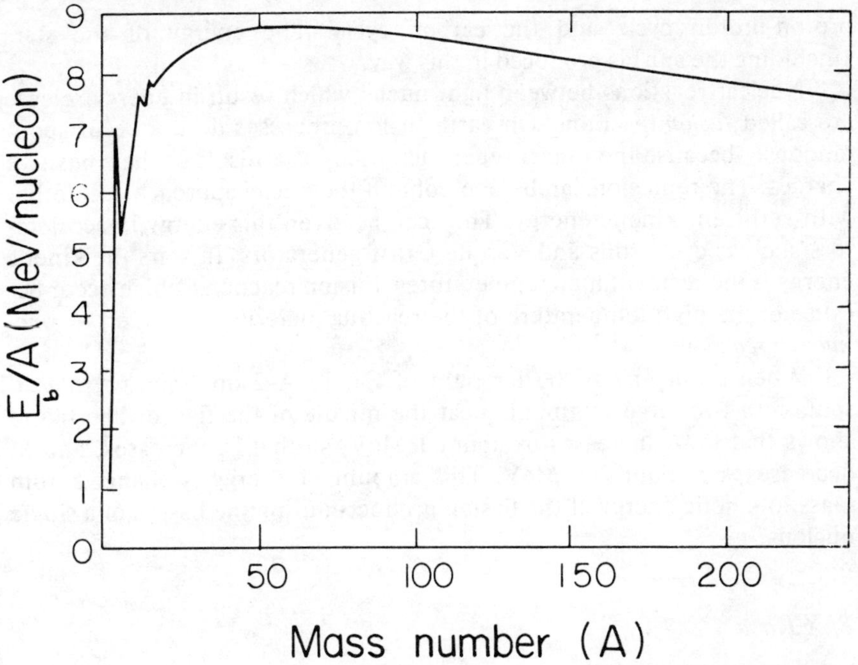

Figure A-2.

Figure A-2 shows a graph of E_b/A as a function of A for the naturally occurring nuclides. There are several interesting features of this diagram:

1. To a first rough approximation the graph shows that the value of E_b/A does not vary much for the natural nuclides.

2. As we look more closely we see that E_b/A decreases from its maximum value of about 8.5 MeV near iron to about 7.5 MeV for the heaviest nuclides.

3. To the left of the maximum the curve also decreases. It has a sharp peak at $A=4$ which represents the α particle, and smaller peaks at ^{12}C and ^{16}O, which can be regarded as multiples of the α particle. For $A=1$ the curve must, of course, descend to zero.

Note that the larger E_b/A is, the more energy is required to separate the A nucleons. This means that greater values of E_b/A correspond to more tightly bound nuclei.

In all radioactive decay the quantity E_b/A increases, as the nucleons assume a configuration in which they are more tightly bound.

7. Fusion and Fission

If an alpha particle is made from its constituents, an amount of energy equal to the binding energy (28 MeV) is changed from mass to energy. This can occur through two sequences of nuclear reactions, called the proton-proton cycle and the carbon cycle. The energy of the stars (including the sun) is produced in this way.

Nuclear reactions between light nuclei which result in energy release are called *fusion* reactions. On earth fusion processes do not occur spontaneously because the nuclei repel each other as a result of their positive charges. The repulsion can be overcome if the nuclei approach each other with sufficient kinetic energy. They can be given this energy in accelerators such as cyclotrons and Van de Graaf generators. In stars the kinetic energy is the result of high temperatures. Fusion reactions which occur because of the high temperature of the reacting materials are called *thermonuclear reactions.*

When a nucleus at the far right of Figure A-2 undergoes *fission,* it splits into two large fragments near the middle of the figure. The figure shows that E_b/A increases by about 1 MeV, so that E_b increases, and M decreases, by about 200 MeV. This amount of energy is changed from mass to kinetic energy of the fission products during the fission of a single nucleus.

8. X Rays

There are two kinds of x rays: characteristic x rays and continuous x rays.

Characteristic x rays are energetic photons emitted in transitions between different energy levels of atoms.

Continuous x rays are photons which accompany the deceleration of electrons (or other charged particles). If, for example, electrons with kinetic energies of 5000 eV are stopped, continuous x rays with energies from zero to 5000 eV will be produced. This process is called *Bremsstrahlung* (from the German for "deceleration radiation").

In an x-ray tube electrons are accelerated to several thousand electron volts. They then hit a target from which both kinds of x rays are emitted.

9. Slowing Down of Charged Particles. Excitation and Ionization

As a fast charged particle passes through matter, there is an electric force between it and the electrons of the atoms which it passes. The force on the electrons causes them to be dislodged so that the atoms either undergo *excitation* to higher energy levels, or, if the electrons leave the atom, *ionization.*

These are the fundamental processes which are utilized in all methods of detection and counting of particles. They are also essential steps in the transformation of the energy which is released in nuclear reactions to heat energy.

Each ionization results in one ion pair, i.e. one electron and one ion. The amount of energy which is required depends on the material, but does not vary much from one material to another. In air (at standard temperature and pressure) one ion pair is produced, on the average, for each 34 eV of energy loss. A one-MeV particle will therefore produce 3×10^4 ion pairs as it loses its energy.

The ions are also charged particles. They are slowed down in the same way and share their energy with the surrounding material. The random motion of the material's atoms and molecules increases. In other words the internal energy increases by the amount of energy which the original charged particle loses. The rise in internal energy corresponds to an increase in the material's temperature.

The processes of excitation and ionization are also responsible for the major chemical and biological effects of radiation. As energy is transferred by these processes, the forces which hold atoms together in molecules are affected so that chemical changes can occur. The forces are particularly delicately balanced in the large and complicated molecules of biological materials, and these are therefore particularly sensitive to damage by radiation.

10. Energy Loss of Photons

Energetic photons lose their energy mainly by transferring their energy to electrons. In the *photoelectric effect* the photon's energy is used in part to remove the electron. The remaining energy is given to the electron as kinetic energy.

In the *Compton effect* only part of the photon's energy is given to an electron. With the rest a photon of lower energy is created.

In *pair production* an electron-positron pair is created. The photon must have at least enough energy to create the pair. Since the electron and positron each have a rest mass equivalent to 0.51 MeV the photon must have at least 1.02 MeV. The remaining energy is given to the newly created pair as kinetic energy.

The electrons now have some or all of the photon's energy. They will lose that energy by excitation and ionization, just like all charged particles.

It must be noted that photons, including x rays and gamma rays, do not ionize directly. They first give their energy to electrons through the photoelectric effect, Compton effect, or pair production. It is the electrons which get their energy through one of these processes which then lose it through excitation and ionization. In other words energetic photons cause ionization indirectly, and are therefore included, together with charged particles, in the term *ionizing radiation.*

11. Ionization: Exposure and the Roentgen

The amount of ionization can be measured by determining the amount of positive and negative charge which is produced. A unit which is often used is the *roentgen* (R). It corresponds to 2.58×10^{-4} coulombs liberated as electrons by ionization in one kg of air at standard temperature and pressure. The number of roentgens is called the *exposure.*

12. Energy Absorbed per Unit Mass; Absorbed Dose; the Gray and the Rad

The amount of energy which is absorbed per unit mass when radiation passes through matter is called the *absorbed dose.* It may be measured in joules/kg. The J/kg is also called the *gray* (Gy). The *rad* is equal to 10^{-2} Gy.

13. Biological Effect: Dose Equivalent; the Sievert and the Rem; the Quality Factor

Different kinds of radiation cause different amounts of biological damage even when the absorbed energy is the same. A quantity called the *dose equivalent* is therefore introduced so as to be the same as the absorbed dose for electrons with energies above 0.03 MeV and for photons. For other kinds of radiation the dose equivalent is different in such a way that it is (at least approximately) proportional to the amount of biological damage which the radiation causes.

The unit for dose equivalent is the *sievert* (Sv). The *rem* (**r**adiation **e**quivalent, **m**an) is equal to 10^{-2} Sv.

The dose equivalent is equal to the absorbed dose times the *quality factor* (Q). The quality factor is a dimensionless factor chosen for different kinds of radiation so as to yield the relation between absorbed dose and dose equivalent. If H is the dose equivalent and D is the absorbed dose, then $H = DQ$.

For electrons with energies greater than 0.03 MeV and for photons the quality factor is equal to one. Other values are as follows:

electrons and β rays with a maximum energy less than 0.03 Mev	1.7
neutrons and protons up to 10 MeV	10.0
naturally occurring alpha particles	10.0
heavy nuclei	20.0

14. Relative Biological Effect

The measurement of actual biological damage is imprecise, and subject to many variables which are difficult to control. The biological effect of a given amount and kind of radiation may depend on the particular organism and the particular part of it which is irradiated, as well as on the particular kind of damage which is considered. It may also vary with the organism's age, its sex, and its general state of health.

Biological experiments, with all their inherent imprecision, are required to compare the biological effects of different kinds of radiation. They show that the effects of electrons with energies greater than 0.03 MeV and of all photons are very closely the same. These are therefore assigned a relative biological effectiveness (RBE) of one. Other radiations, in general, cause greater damage, and biological experiments lead to values of the appropriate RBE, equal to the ratio of the damage for the examined radiation to the damage caused by radiation with an RBE of one.

It must be emphasized that the RBE may, as has been indicated earlier, depend on factors other than just the particular kind of radiation, and may, in general, vary from one experiment to the other.

It is because of the uncertainty of the RBE that the quality factor Q was introduced. The quality factor is a definite number, assigned by an appropriate committee, and accepted as the value to be used to relate the units for absorbed dose (the gray and the rad) to the units of dose equivalent (the sievert and the rem).

The procedure is, to some extent, analogous to the establishment of an "accepted" value for the speed of light, or for Avogadro's number, by a committee, based on the best available information. If new experiments are made the accepted value may be changed. The relation between the radiation units is more complicated because the absorbed dose is a definite physical quantity (absorbed energy per unit mass), while the dose equivalent is related to biological effects which are much more difficult to quantify.

The term quality factor has come into use only recently. Earlier the term "relative biological effectiveness" was used both in the sense in which it is defined here, and for the quantity which is now called the quality factor. The two terms are still frequently used interchangeably, especially in popular articles in newspapers and magazines. Normally there is no ambiguity about the meaning.

15. Neutron Interactions

Neutrons are particularly efficient in initiating nuclear reactions because they are not repelled by the electric charge of nuclei as are protons and alpha particles.

Nuclear reactions may result in the production of radioactive nuclides, and so cause "induced radioactivity." Among radiations of a few MeV or less, only neutrons can do so to any important extent. Protons, alpha-particles and other nuclei in this energy range cannot normally come sufficiently close to the target nuclei to cause nuclear reactions because of the electric repulsion. Photons and electrons also do not cause nuclear reactions at these energies.

16. Slow-Neutron Reactors

Only about three quarters of one percent of natural uranium is the isotope ^{235}U. Its importance stems from the fact that even a neutron with vanishingly small kinetic energy can initiate fission in a ^{235}U nucleus. Moreover the probability of fission goes up as the neutron slows down. ^{235}U is therefore one of a few nuclides which can be used as fuel in *slow-neutron reactors.*

Fission is accompanied by the emission of neutrons whose energy is several MeV. In a slow-neutron reactor these neutrons are slowed down by a *moderator* to increase their chances for initiating further fission events. The moderator may be ordinary (light) water, heavy water (deuterium oxide) or graphite. The fuel may be ^{235}U in natural uranium, or in uranium "enriched" to a higher concentration of ^{235}U.

Energy is released by fission, mostly as kinetic energy of the fission fragments. The fission fragments are large nuclei, consisting of a number of nucleons of order of magnitude 10^2. They slow down by causing excitation and ionization, and, like other charged particles, eventually share their kinetic energy with the surrounding atoms and molecules as heat.

The heat is carried away by a cooling medium, usually water, and can then be used in a *heat engine* to produce mechanical energy. The mechanical energy, in turn, is converted to electrical energy in an electric generator.

The rate at which fission occurs, and at which heat is therefore produced, can be regulated by *control rods.* These are made of a material which absorbs neutrons strongly. The introduction of a control rod into a reactor removes neutrons and reduces the heat output. Boron and cadmium are elements which absorb neutrons particularly strongly, and are therefore used to reduce the number of neutrons in nuclear reactors.

17. Breeder Reactors

A neutron may be absorbed by a ^{238}U nucleus, forming ^{239}U. Successive beta decay leads through neptunium (Z=93) to plutonium (Z=94). Here is the sequence of reactions:

$$^{238}U + n \rightarrow {}^{239}U + \gamma$$

$$^{239}U \rightarrow {}^{239}Np + \beta^- + \bar{\nu} \qquad (T_{1/2} = 23.5 \text{ min.})$$

$$^{239}Np \rightarrow {}^{239}Pu + \beta^- + \bar{\nu} \qquad (T_{1/2} = 2.35 \text{ days})$$

$$^{239}Pu \rightarrow {}^{235}U + \alpha \qquad (T_{1/2} = 2.44 \times 10^4 \text{ years})$$

The ^{239}Pu is, like ^{235}U, fissionable by slow neutrons. It is produced in all reactors which contain ^{238}U, but a reactor built especially for plutonium production is called a breeder reactor. It makes the much more abundant uranium isotope ^{238}U accessible as a fuel.

Problem Set A-3

1. The mass of a neutral atom of ^{12}C is exactly 12 u. The mass of one mole is 12 g. Use these facts to calculate the number of kg equal to 1 u.
2. One kg is equivalent to 9×10^{16} J. From the conversion factors between eV and J, and between u and kg, find the number of J and of MeV equivalent to 1 u.
3. From the masses in Table A-2, calculate E_b and E_b/A for $^{4}_{2}He$, $^{56}_{26}Fe$, and $^{239}_{92}U$. (Check your results on Fig. A-2.)
4. A charged particle, whose initial energy is 1 MeV, is slowed down, and finally stopped, in air. Find the following:
   (a) The number of electrons which are liberated by ionization.
   (b) The amount of charge (in Coulombs) which is liberated as electrons.
   (c) The answer to part b gives the amount of charge (in C) liberated when an energy of 1 MeV is absorbed. Use this result to find the amount of energy which is absorbed when 2.58×10^{-4} C are liberated. (Calculate this energy both in MeV and in J.)
   (d) Use your answer to part c to find the number of Gy and the number of rads equal to the energy absorbed per unit mass of air when the exposure is 1 R.

Table A-1: Conversion Factors, Constants, & Definitions

Conversion factors

$1\ \text{eV} = 1.6 \times 10^{-19}\text{J}$	(electron volt to joule)
$1\ \text{keV} = 10^{3}\ \text{eV}$	
$1\ \text{MeV} = 10^{6}\ \text{eV}$	
$1\ \text{u} = 931.5\ \text{MeV/c}^2$	(unified atomic mass unit to Mev)
$1\ \text{u} = 1.66 \times 10^{-27}\text{kg}$	(unified atomic mass unit to kg)
$1\ \text{Ci} = 3.7 \times 10^{10}\text{Bq}$	(curie to becquerel)
$1\ \mu\text{Ci} = 10^{-6}\ \text{Ci}$	
$1\ \text{rad} = 10^{-2}\ \text{Gy}$	(rad to gray)
$1\ \text{mrad} = 10^{-3}\ \text{rad}$	
$1\ \text{rem} = 10^{-2}\ \text{Sv}$	(rem to sievert)
$1\ \text{mrem} = 10^{-3}\ \text{rem}$	
$1\ \text{yr} = 3.16 \times 10^{7}\ \text{s}$	(year to second)
$1\ \text{T} = 10^{4}\ \text{G}$	(tesla to gauss)

Constants

speed of light	$c = 3.00 \times 10^{8}\ \text{m/s}$
electronic charge	$e = 1.60 \times 10^{-19}\ \text{C}$
Planck's constant	$h = 6.63 \times 10^{-34}\ \text{Js}$
Boltzmann's constant........	$k = 1.38 \times 10^{-23}\ \text{J/K}$
Avogadro's number	$N_o = 6.02 \times 10^{23}\ \text{mol}^{-1}$
Coulomb force constant	$K = 1/4\pi\epsilon_o = 9.00 \times 10^{9}\text{N m}^2/\text{C}^2$
magnetic force constant	$\mu_o/4\pi = 10^{-7}\ \text{N/A}^2$
universal gas constant	$R = 8.31\ \text{J/K mol}$

Definitions, etc.

1 Bq = 1 disintegration/second

1 rad = 10^{-2} J/kg

1 R = 2.58×10^{-4} C/kg of electrons in air at STP

For an element whose atomic mass is M there are N_A atoms in M grams. M is approximately (but not exactly!) equal to the mass number A.

For a kinetic temperature, kT, of 1 eV, T - 1.16×10^4 K.

The average energy lost by a charged particle when it produces one ion pair in air at STP is 34 eV.

Relation between decay constant (λ) and half-life ($T_{1/2}$):
$\lambda T_{1/2} = 0.693$

Table A-2

Masses of Elementary particles

$m_e = 9.11 \times 10^{-31}$ kg $= 5.49 \times 10^{-4}$ u — $m_e c^2 = 0.511$ MeV

$m_p = 1.6725 \times 10^{-27}$ kg $= 1.00728$ u — $m_p c^2 = 938.3$ MeV

$m_n = 1.6748 \times 10^{-27}$ kg $= 1.00866$ u — $m_n c^2 = 939.6$ MeV

Some Atomic Masses

(these are the masses of the neutral atoms in unified atomic mass units)

$^{1}_{0}n$	1.00866	$^{12}_{6}C$	12.00000
$^{1}_{1}H$	1.00783	$^{14}_{6}C$	14.00324
$^{2}_{1}H$	2.01410	$^{14}_{7}N$	14.00307
$^{3}_{1}H$	3.01605	$^{46}_{26}Fe$	45.93494
$^{3}_{2}He$	3.01603	$^{231}_{90}Th$	231.03630
$^{4}_{2}He$	4.00260	$^{235}_{92}U$	235.04392
$^{6}_{3}Li$	6.01513	$^{238}_{92}U$	238.05077
$^{7}_{3}Li$	7.01600		

Index to the Appendix

Index to the Appendix, concluded